INTRODUCTION TO PROBABILITY WITH TEXAS HOLD'EM EXAMPLES

EXAMPLES

SECOND EDITION

SECOND EDITION

INTRODUCTION TO PROBABILITY WITH TEXAS HOLD'EM EXAMPLES

EXAMPLES

FREDERIC PAIK SCHOENBERG

CRC Press
Taylor & Francis Group
Boca Raton London New York

CRC Press is an imprint of the
Taylor & Francis Group, an **informa** business

A CHAPMAN & HALL BOOK

CRC Press
Taylor & Francis Group
6000 Broken Sound Parkway NW, Suite 300
Boca Raton, FL 33487-2742

© 2017 by Taylor & Francis Group, LLC
CRC Press is an imprint of Taylor & Francis Group, an Informa business

No claim to original U.S. Government works

Printed on acid-free paper
Version Date: 20161021

International Standard Book Number-13: 978-1-4987-7618-9 (Paperback)

Library of Congress Cataloging-in-Publication Data

Names: Schoenberg, Frederic Paik.
Title: Introduction to probability with Texas hold'em examples / Frederic Paik Schoenberg.
Description: Second edition. | Boca Raton : Taylor & Francis, 2017. | "A CRC title." | Includes bibliographical references and index.
Identifiers: LCCN 2016026388 | ISBN 9781498776189 (hardback)
Subjects: LCSH: Probabilities--Textbooks. | Poker--Miscellanea.
Classification: LCC QA273 .S377 2017 | DDC 519.2--dc23
LC record available at https://lccn.loc.gov/2016026388

Visit the Taylor & Francis Web site at
http://www.taylorandfrancis.com

and the CRC Press Web site at
http://www.crcpress.com

Contents

Preface to the First Edition   ix
Preface to the Second Edition   xv

Chapter 1   Probability Basics   1

    1.1   Meaning of *Probability*   1
    1.2   Basic Terminology   3
    1.3   Axioms of Probability   4
    1.4   Venn Diagrams   5
    1.5   General Addition Rule   7
    Exercises   12

Chapter 2   Counting Problems   15

    2.1   Sample Spaces with Equally
        Probable Events   16
    2.2   Multiplicative Counting Rule   20
    2.3   Permutations   22
    2.4   Combinations   26
    Exercises   49

Chapter 3   Conditional Probability and Independence   57

    3.1   Conditional Probability   57
    3.2   Independence   64
    3.3   Multiplication Rules   66
    3.4   Bayes' Rule and Structured Hand Analysis   71
    Exercises   77

Chapter 4   Expected Value and Variance   83

    4.1   Cumulative Distribution Function
        and Probability Mass Function   83
    4.2   Expected Value   85
    4.3   Pot Odds   92
    4.4   Luck and Skill in Texas Hold'em   104

4.5   Variance and Standard Deviation   119
4.6   Markov and Chebyshev Inequalities   123
4.7   Moment-Generating Functions   125
Exercises   127

Chapter 5   Discrete Random Variables   137

5.1   Bernoulli Random Variables   138
5.2   Binomial Random Variables   140
5.3   Geometric Random Variables   143
5.4   Negative Binomial Random Variables   146
5.5   Poisson Random Variables   147
Exercises   152

Chapter 6   Continuous Random Variables   157

6.1   Probability Density Functions   157
6.2   Expected Value, Variance, and
        Standard Deviation   161
6.3   Uniform Random Variables   164
6.4   Exponential Random Variables   174
6.5   Normal Random Variables   177
6.6   Pareto Random Variables   181
6.7   Continuous Prior and
        Posterior Distributions   184
Exercises   187

Chapter 7   Collections of Random Variables   191

7.1   Expected Value and Variance of Sums
        of Random Variables   191
7.2   Conditional Expectation   197
7.3   Law of Large Numbers and the
        Fundamental Theorem of Poker   200
7.4   Central Limit Theorem   208
7.5   Confidence Intervals for the
        Sample Mean   215
7.6   Random Walks and the Probability
        of Ruin   220
Exercises   233

Chapter 8 Simulation and Approximation
Using Computers   239

Exercises   249

Appendix A: Abbreviated Rules of Texas Hold'em   253

Appendix B: Glossary of Poker Terms   257

**Appendix C: Solutions to Selected
Odd-Numbered Exercises**   261

References and Suggested Reading   269

Index   273

Preface to the First Edition

I am a lousy poker player. Let me get that out of the way right off the bat. If you are reading this book in the hope that you will learn strategy tips on how to be a better poker player, you are bound to be disappointed. This is not a book on how to use probability to play Texas Hold'em. It is a textbook using Texas Hold'em examples to teach probability.

The other thing I want to state right from the outset is that I in no way intend this book to be an endorsement of gambling. Poker, like all forms of gambling, can be addictive and dangerous. The morality of gambling has been properly questioned by many for a host of reasons, and among them is the fact that many people, especially those who can least afford it, often lose more than they should prudently risk. The rise of online gambling recently, especially among students at colleges and universities, is cause for serious concern. When I have taught my course on poker and probability at UCLA in the past, I have always started out on the first day by lecturing about the dangers of gambling, and my first required readings for the students are handouts on the perils of gambling addiction.

The purpose of this book is not to promote gambling or to teach students how to play poker. Instead, my intention is to use students' natural interest in poker to motivate them to learn important topics in probability. The first few times I taught probability, I was disappointed by many of the examples in the books. They typically involved socks in sock drawers or balls in urns. Most of my students did not even know what an urn was, and they certainly were not motivated when informed of its meaning. I thought it would be interesting to try to teach the same topics as those covered in most probability texts using only examples from poker. I was happy to find that students seemed to vastly prefer these examples and that it was hardly a challenge at all to motivate even quite complex subjects using poker. In fact, I needed to look no further than Texas Hold'em, which is currently the most popular poker game, to illustrate all the standard undergraduate probability topics and even some more advanced topics. While some have urged me to discuss other poker games, I have decided instead to stick exclusively to Texas Hold'em examples for two reasons. The first is that Texas Hold'em is more popular and more commonly televised than other games and hence may be of greater interest to students. The second reason is simply brevity. The purpose of this book is to teach probability, not the rules and intricacies of various poker games, and I found absolutely no reason to search beyond Texas Hold'em to illustrate any probability topic.

The topics covered in this book are similar to those in most undergraduate probability textbooks with a few exceptions. I have added sections on special topics, including a few specialized poker issues, such as the quantification of luck and skill in Texas Hold'em, and some topics typically found in graduate probability texts, such as the ballot theorem and the arcsine law.

I will doubtless be criticized by some of my colleagues for writing this book, because poker is not only perceived as immoral but also as frivolous. While many probabilists and statisticians might feel that probability should be taught using more serious, scientific examples, I disagree. I fully acknowledge the downsides of poker, but poker has its good qualities as well. Texas Hold'em is fun, and its current popularity can be used to attract students and keep them interested. Texas Hold'em involves a blend of luck and skill that may be extremely frustrating at times for players, but can also be wonderfully intriguing, and incidentally is in many ways similar to other pursuits in life that seem to rely on a similar blend of skill and fortune, such as a search for a job or for love. Most importantly, in my opinion, Texas Hold'em is at its heart an intellectual pursuit. Gambling games such as poker have inspired many of the most important ideas in probability theory, including Bayes' theorem and the law of large numbers that have found applications in so many scientific disciplines.

Aside from using exclusively Texas Hold'em examples, one other feature that I hope may make this book unique as a probability textbook is that I have tried, wherever possible, to use only real examples—not *realistic*, but real examples from actual hands of Texas Hold'em shown played in the World Series of Poker or other major tournaments or televised games. The search for these hands was time-consuming but enjoyable, and the use of real examples may help to keep students interested. Sometimes the probability topic discussed is somewhat tangent to the main issue that makes a poker hand interesting, but hopefully readers can look past this. A list of links to some of the hands referred to in this book is at http://www.stat. ucla.edu/~frederic/hands.html.

When I have taught this course in the past, in addition to homework and exams, I assigned the students two computational projects. On the first project, the students were asked to write a function in R that took various inputs, including their cards, the betting before them, their number of chips, the number of players at the table, and the size of the blinds, and then output a bet size of 0 or their number of chips. That is, they had to write a program to fold or go all-in. I would then have their computer programs compete in tournaments that I ran multiple times. For the final project, they were asked to write a Texas Hold'em program in R that could be more complicated and did not require them to go all-in or fold, but allowed them to bet intermediate amounts. Some students sincerely enjoyed these projects and wrote quite elaborate functions. Several students felt that this was their favorite aspect of the course. I have compiled functions for instructors to use to run these tournaments, as well as some examples of the students' functions, into a public *R* package called *holdem*, which may be freely downloaded from www.r-project.org. Some description is given in Chapter 8 as well.

I have a lot of people to thank. First and foremost, I thank my wife, Jean, not only for supporting me throughout the writing of this book but also for indirectly introducing me to Texas Hold'em by taking me on a surprise trip to Las Vegas for my birthday several years ago. Gamma, Dad, Mom, Randy, Marlena, and Melanie not only gave me endless support but also played tons of cards with me growing up, Aidee's help made writing the book possible, and Bella gave me inspiration and emotional support. My friend Craig Berger taught me the ins and outs of poker strategy, and David Diez taught me how to make an *R* package. Keith Wilson and Daniel Lawrence provided numerous stimulating conversations about poker, as did Tom Ferguson, Arnulfo Gonzalez, Reza Gholizadeh, and John Fernandez. I want to thank Jamie Gold, who came and spoke to my class and was extremely nice and entertaining. I am also grateful for

Nine-month-old poker players Max (left) and Gemma (right)
Paik Schoenberg, May 2010.

the excellent probability texts by Feller (1966, 1967), Billingsley (1990), Pitman (1993), Ross (2009), and Durrett (2010), from which much of the content in this book was taken. My twin children, Gemma and Max, were born while I was writing this book, and provided not only great inspiration but also ample distraction. Any mistakes are their fault!

Preface to the Second Edition

The poker boom continues. In the 5 years since this book was originally published, the popularity of Texas Hold'em has remained extremely strong and is spreading to people from an increasingly wide array of nationalities and demographic backgrounds. We have 5 new World Champions, 5 years of exciting televised tournaments, and 5 more years of incredible poker hands to discuss. In the second edition, I have added numerous examples and exercises involving recent poker hands, especially those from the Main Event of the World Series of Poker. Many corrections, updates, and editing improvements have also been made, making this edition much easier and more fun to read.

The response to the first edition, and to its accompanying course, which I have taught every year at UCLA, was quite different than I imagined. I knew the students would like it, but I expected to receive some criticism from the statistical and academic community for using only poker examples. Fortunately, none arrived. Honestly, I was worried some would think talking about Texas Hold'em in an undergraduate course would be seen as too much fun, but I have been very pleasantly surprised with my colleagues and their open-mindedness and willingness to accept that poker is at its core an intellectual pursuit and one that can be used to inspire students to learn the fundamentals of probability.

I also thought there might be more reaction to my proposed quantification of luck and skill in Section 4.4. There was instead, however, some issue taken with a particular paragraph in Section 7.6, where I express a bit of criticism for a portion of Chapter 22 of Chen and Ankenman's book *The Mathematics of Poker*. Some discussion about this appeared on AndrewGelman.com in August 2014. I have decided to keep my paragraph unchanged in the current edition, and readers are invited to read both texts and see what they think. I would be interested to hear from poker players whether they feel the approximation in Chen and Ankenman is useful for them in practice, and I would be greatly interested in reactions to my proposed definitions of *luck* and *skill*, the terms so many articles and texts on game theory seem to use without providing any quantitative definitions.

It has been a great delight for me to use this book to teach at UCLA, and I thank all the students for going on this journey with me. I have had several students who were not interested in probability and statistics before taking this course and subsequently became statistics majors. I have had students who did not know how many cards were in a deck when the course began, and other students who had already enjoyed lucrative careers as professional poker players. Maybe someday I will turn on the TV and see one of my former students holding the Main Event bracelet. Still, this is a probability book, not a poker strategy book, so perhaps my goal instead should be to see one of my former students hoisting the Fields Medal or becoming a famous statistician.

As before, I thank my family, especially my wife and kids, and I also want to thank David Grubbs, Amber Donley, a very helpful anonymous reviewer, and the rest of the team at CRC Press/Taylor & Francis Group, as well as Bruce McCullough for all their help and support.

CHAPTER 1

Probability Basics

1.1 Meaning of *Probability*

It's hand 229 of day 7 of the 2006 World Series of Poker (WSOP) Main Event, the biggest poker tournament ever played, and after the elimination of 8770 entrants it is now down to the final three players. The winner gets a cash prize of $12 million, second place just over $6.1 million, and third place gets about $4.1 million. Jamie Gold is the chip leader with $60 million in chips, Paul Wasicka has $18 million, and Michael Binger has $11 million. The blinds are $200,000 and $400,000, with $50,000 in antes. (Note: For readers unfamiliar with Texas Hold'em, please consult Appendices A and B for a brief explanation of the game and related terminology used in this book.) Gold calls with 4♠ 3♣, Wasicka calls with 8♠ 7♠, Binger raises with A♥ 10♥, and Gold and Wasicka call. The flop is 10♣ 6♠ 5♠. Wasicka checks, perhaps expecting to raise, but by the time it gets back to him, Binger has bet $3.5 million and Gold has raised all-in. What would you do, if you were Paul Wasicka in this situation?

There are, of course, so many issues to consider. One relatively simple probability question that may arise is this: *given all the players' cards and the flop, if Wasicka calls, what is the probability that he will make a flush or a straight?*

This is the type of calculation we will address in this book, but before we get to the calculation, it is worth examining the question a bit. What does it mean to say that the probability of some event, like Wasicka's making a flush or a straight, is, say, 55%? It may surprise some readers to learn that considerable disagreement continues among probabilists and statisticians about the definition of the term *probability*. There are two main schools of thought.

The *frequentist* definition is that if one were to record observations under the exact same conditions over and over, with each observation independent of the rest, then the event in question would ultimately occur 55% of the time. In other words, to say that the probability of Wasicka making a flush or a straight is 55% means that if we were to imagine *repeatedly* observing situations just like this one, or if we were to imagine dealing the turn and river cards repeatedly, each time reshuffling the remaining 43 cards before dealing, then Wasicka would make a flush or a straight 55% of the time.

The *Bayesian* definition is that the quantity 55% reflects one's subjective feeling about how likely the event is to occur: in this case, because the number is 55%, the feeling is that the event is slightly more likely to occur than it is not to occur.

The two definitions suggest very different scientific questions, statistical procedures, and interpretations of results. For instance, a Bayesian may discuss the probability that life exists on Mars, while a frequentist may argue that such a construct does not even make sense. Frequentists and Bayesians have had spirited debates for decades on the meaning of probability, and consensus appears to be nowhere in sight.

While experts may differ about the meaning of probability, they have achieved unanimous agreement about the *mathematics* of probability. Both frequentists and Bayesians agree on the basic rules of probability—known as the *axioms of probability*. These three axioms are discussed in Section 1.3, and the methods for calculating probabilities follow from these three simple rules. There is also agreement about the proper notation for probabilities: we write $P(A)$ = 55%, for instance, to connote that the probability of event A is 55%.

1.2 Basic Terminology

Before we get to the axioms of probability, a few terms must be clarified. One is the word *or*. In the italicized question about probability in Section 1.1, it is unclear whether we mean the probability that Wasicka makes either a flush or a straight *but not both* or the probability that Wasicka makes a flush or a straight *or both*. English is ambiguous regarding the use of *or*. Mathematicians prefer clarity over ambiguity and have agreed on the convention that A *or* B always means A *or* B *or both*. If one means A *or* B *but not both*, one must explicitly include the phrase *but not both*.

Of course, in some situations, A and B cannot both occur. For instance, if we were to ask what the probability is that Wasicka makes a flush or three 8s on this hand, it is obvious that they cannot both happen: if the turn and river are both 8s, then neither the turn nor the river can be a spade, since Wasicka already has 8♠. If two events A and B cannot both occur, i.e., if $P(A \text{ and } B)$ = 0, then we say that the events are *mutually exclusive*. We usually use the notation AB to denote the event A *and* B, so the condition for mutual exclusivity can be written simply $P(AB)$ = 0.

The collection of all possible outcomes is sometimes called the *sample space*, and an *event* is a subset of elements of the sample space. For instance, in the WSOP example described in the beginning of Section 1.1, if one were to consider the possibilities that might come on the turn, one might consider the sample space to be all 52 cards. Of course, if we know what cards have been dealt to the three players and which cards appeared on the flop, then these nine cards have zero probability of appearing on the turn, and we may consider the remaining 43 cards equally likely. The event that the turn is the 7♦ is an event consisting of a single element of the sample space, and the event that the turn is a diamond is an event consisting of 13 elements.

Given an event A, we use the notation A^c to mean the *complement* of A, or in other words the event that A does not occur. If A is the event that the turn is a diamond, for instance, then A^c is the event that the turn is a club, heart, or spade. For any event A, the events A and A^c are always mutually exclusive and *exhaustive*, meaning that together they cover the entire sample space.

1.3 Axioms of Probability

Three basic rules or *axioms* must govern all probabilities:

1. $P(A) \geq 0$.
2. $P(A) + P(A^c) = 1$.
3. If $A_1, A_2, A_3, \ldots$ are mutually exclusive events, then $P(A_1 \text{ or } A_2 \text{ or } \ldots A_n) = P(A_1) + P(A_2) + \ldots + P(A_n)$,

where n may be a positive integer or ∞.

Axiom 1 says that all probabilities must be at least 0, and because of this and axiom 2, no probability can be greater than 1. According to axiom 2, if the probability that A will not occur, is, say, 45%, then we know that the probability of A occurring is 55%. It is sometimes convenient to compute the probability of A *not* occurring en route to computing the probability of A. Axiom 3 is sometimes called the *addition rule for mutually exclusive events*. Note that it is usually rather obvious when events are mutually exclusive. For example, if you play a single hand of Texas Hold'em, A_1 might be the event that you are dealt pocket aces and A_2 the event that you are dealt pocket kings. They cannot both occur, so these two events are mutually exclusive, and according to axiom 3 the probability of your being dealt either pocket aces or pocket kings on this single hand is equal to the probability of your being dealt pocket aces plus the probability of your being dealt pocket kings. This may seem obvious, but it is worth noting that the fact that we may add these two probabilities together to find the probability that one will occur is not something that can be proven; it is a fundamental principle we agree to *assume* when discussing probabilities.

Note also that axiom 3 does not apply directly to the question discussed in Section 1.1 involving the probability of Wasicka making a flush or a straight. It is possible (if, for instance, the turn and river are the 9♣ and Q♠) for both to occur, so the event that Wasicka makes a straight and the event that he makes a flush are not mutually exclusive.

A final comment worth noting about the axioms of probability is that they imply that if there are n equally likely events, *exactly one of which must occur*, then the probability associated with any one of the events is simply $1/n$. This may again seem obvious, but note that it can be proven directly from axiom 3 (see Exercise 1.3). This fact is important for many questions involving Texas Hold'em, because in many cases the outcomes are equally likely. The subject of equally likely events is taken up further in Section 2.1.

1.4 Venn Diagrams

Most readers have likely seen Venn diagrams, so they will be reviewed only briefly here. One question that is sometimes overlooked in discussing Venn diagrams is, *why* can one use shapes such as those shown in Figure 1.1 to address probability questions?

The answer has everything to do with the axioms of probability discussed in Section 1.3. All the rules of probability discussed in this book follow from these three simple axioms. Note that they apply not only

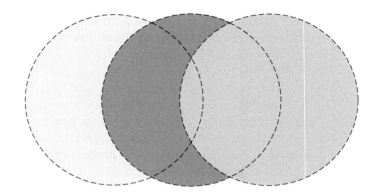

FIGURE 1.1   Venn diagram.

to probabilities but also to other constructs, including areas. Suppose you have a piece of paper of area 1, and that A_1, A_2, A_3, ... represent shapes on the paper. Suppose also that $P(A)$ means the *area* occupied by shape A (instead of probability); A^c is the portion of the paper not occupied by shape A; AB denotes the intersection of A and B, i.e., the portion of the paper occupied by both A and B; and A *or* B indicates the union of A and B, i.e., the portion of the paper occupied by shapes A or B or their intersection. Then the three axioms would still be true: any imaginable shape on the paper would necessarily have an area greater than or equal to 0; the area covered by the shape plus the area not covered by the shape would equal 1; and shapes would be mutually exclusive if their area of intersection were 0. So if shapes A_1, A_2, A_3, ... are mutually exclusive, then obviously the total area covered by them is simply the area of A_1 + the area of A_2 + the area of A_3 + ...

Thus, the same axioms that govern probability also apply to areas on a paper of total area 1. We will see the connection between probability and area several times in this book. It is worth noting that in further studies of probability at the graduate and research levels, the connections are made very explicit between probability and measure theory, the latter of which is essentially the study of principles related to areas, volumes, and other means of measuring shapes or other objects.

1.5 General Addition Rule

Axiom 3 in Section 1.3 is sometimes called the *addition rule for mutually exclusive events.* How does one compute the corresponding probability for events that may *not* be mutually exclusive? In general, for any events A and B,

$$P(A \text{ or } B) = P(A) + P(B) - P(AB).$$

The above may be called the *general addition rule* because it applies to events A and B whether they are mutually exclusive or not. Of course, when A and B are mutually exclusive, the last term, $P(AB)$, is equal to 0. The general addition rule follows directly from the axioms of probability and may be demonstrated quite easily by breaking up $A \text{ or } B$ into mutually exclusive events. That is, observe that

$$A \text{ or } B = AB^c \text{ or } AB \text{ or } A^c B,$$

and that these three events, AB^c, AB, and $A^c B$, are mutually exclusive. Therefore,

$$P(A \text{ or } B) = P(AB^c \text{ or } AB \text{ or } A^c B)$$
$$= P(AB^c) + P(AB) + P(A^c B).$$

One can also form the partition $A = AB^c \text{ or } AB$, and since AB^c and AB are mutually exclusive, it follows from axiom 3 that $P(A) = P(AB^c) + P(AB)$. Similarly, $B = A^c B$ or AB, so $P(B) = P(A^c B) + P(AB)$. Thus,

$$P(A) + P(B) - P(AB) = [P(AB^c) + P(AB)] + [P(A^c B) + P(AB)] - P(AB)$$

$$= P(AB^c) + P(AB) + P(A^cB)$$

$$= P(A \text{ or } B).$$

The following five examples refer to the WSOP hand discussed previously. For each example, assume we have only the information given in Section 1.1: that Wasicka had 8♠ 7♠, Binger had A♥ 10♥, Gold had 4♠ 3♣, and the flop was 10♣ 6♠ 5♠. Assume that the remaining 43 cards are all equally likely to appear as turn or river cards and suppose that Wasicka had called.

Example 1.5.1

What is the probability of Wasicka making a straight flush on the turn?

Answer Wasicka needs the 4♠ or 9♠ to make a straight flush, but Gold has the 4♠, so the only remaining possibility is the 9♠. Since there are 43 equally likely possibilities for the turn, exactly one of which must occur, the probability associated with each of them appearing on the turn must be 1/43, and therefore the probability of the turn being the 9♠ is 1/43.

Example 1.5.2

What is the probability of Wasicka making a straight flush on the turn or river?

Answer Note that, from the definition of the word *or* given in Section 1.2, this question is really asking what the probability is of Wasicka making a straight flush on the turn or river *or both*. However, the only card that can give Wasicka a straight flush is the 9♠, and it cannot appear both on the turn and the river. Thus, if A is the event that the 9♠ comes on the turn and B is the event that the 9♠

comes on the river, then A and B are mutually exclusive, and by axiom 3, the probability of A or B is $P(A) + P(B)$, which is simply 1/43 + 1/43 = 2/43.

Note that $P(B)$ in the answer above is 1/43, not 1/42. It might help to think of $P(B)$ as the *frequency* with which the river will be the 9♠, if we were to repeat this scenario over and over again. Clearly, each of the 43 remaining cards is equally likely to appear on the river, so the probability associated with the 9♠ is simply 1/43.

Example 1.5.3

What is the probability that the turn card does not give Wasicka a straight flush *and* the river card also does not give Wasicka a straight flush?

Answer This is the complement of the event in Example 1.5.2, so by axiom 2, the answer is 1 − 2/43 = 41/43.

Example 1.5.4

What is the probability of Wasicka making a flush on the turn?

Answer Eight spades are left in the deck: A♠, K♠, Q♠, J♠, 10♠, 9♠, 3♠, and 2♠. As noted in Example 1.5.1, each of the 43 remaining cards has a probability of 1/43 of appearing on the turn. Let A_1 be the event that the turn is the A♠, A_2 be the event that the turn is the K♠, and so on. Note that $A_1, A_2, ..., A_8$ are *mutually exclusive*, since it is impossible for more than one of these cards to appear on the turn. Thus by axiom 3,

$$P(A_1 \text{ or } A_2 \text{ or } ... \text{ or } A_8) = P(A_1) + P(A_2) + ... + P(A_8)$$

$$= 1/43 + 1/43 + ... + 1/43$$

$$= 8/43$$

$$\sim 18.6\%.$$

Two simple approximations used frequently by poker players are known as the *Rule of Two* and the *Rule of Four*. If you are behind your opponent presently, we call an *out* a card that will give you the lead. Let x be the number of outs left in the deck. The Rule of Two says that your probability of an out coming on the next card (e.g., the turn) is approximately $2x\%$. The Rule of Four says that your probability of an out coming on the turn or river (or both) is approximately $4x\%$. Of course, the term *out* can be replaced by an element of any collection of x cards. The idea behind the Rule of Two is that, if one is interested in the probability of one of x possible cards coming on the turn, given only one's hole cards and the three flop cards, then there are 47 cards left in the deck, each equally likely to appear, so the probability of one of these x cards coming is $x/47 \sim x/50 = 2x/100$. For the Rule of Four, the idea is that, again given only one's hole cards and the three flop cards,

$$P(out\ on\ turn\ or\ river) = P(out\ on\ turn) + P(out\ on\ river)$$
$$- P(both).$$

If x is small, then this last term, the probability of an out coming on both the turn and river, will be small and is therefore neglected, and the other terms are each equal to $x/47 \sim 2x/100$. Note that $2x/100$ is a slight underestimate of $x/47$, but this difference is offset by the exclusion of the term $P(both)$, resulting in a quite accurate estimate for small x. The approximation becomes poor, however, when x is large, as discussed further in Section 2.4. In Example 1.5.4, $x = 8$, so the Rule of Two yields the approximation 16%, whereas the actual answer was 18.6%. However, in Example 1.5.4 we also assumed the cards of Binger and Gold were known; otherwise, the correct answer would be $8/47 \sim 17.0\%$.

Example 1.5.5

What is the probability of Wasicka making a flush on the turn or the river?

Answer This one is not so simple. If A is the event that the turn is a spade and B is the event that the river is a spade, then A and B are *not* mutually exclusive. We will return to the correct solution to this problem after discussing permutations and combinations, in Example 2.4.3. For now, we may apply the Rule of Four to obtain an approximate answer. Eight spades remain, so $x = 8$, and the Rule of Four yields 32% as the approximate solution.

Example 1.5.6

Mark Newhouse, who became one of the WSOP Main Event's final nine players (called the *November Nine*) in 2013, made the November Nine again in 2014. Of the 6683 entrants in the 2014 Main Event, if each player uses the same strategy, then what is the probability that any one given player would become part of the November Nine?

Answer The given player would be equally likely to wind up in 1st place, 2nd place, ..., or 6683rd place, so the probability of being in the top nine is $9/6683 \sim 0.135\%$.

Example 1.5.7

On day 5 in the 2015 WSOP Main Event, Andrew Moreno went all-in with A♦ K♦ and Dax Greene called with A♠ A♥. The flop came 5♦ K♠ K♥, giving Moreno three of a kind. Suddenly Greene needed the A♣ to come (and the K♣ not to come) to win the hand. Given the players' cards and the three cards on the flop, what is the probability of the A♣ coming on the turn or river?

Answer   Since the A♣ is Greene's only out, the Rule of Four suggests an estimate of 4% for the A♣ to come on the turn or river. Computed exactly,

P(A♣ on turn or river)

  $= P$(A♣ on turn) $+ P$(A♣ on river)

  $- P$(A♣ on both turn and river)

  $= P$(A♣ on turn) $+ P$(A♣ on river)

  $= 1/45 + 1/45 \sim 4.44\%$.

(In the actual hand, the turn and river were the 7♠ and 8♣, so Moreno won the pot.)

Exercises

1.1 Sometimes it matters whose hand is second best. With the Gold, Wasicka, and Binger hands, for instance, if both Wasicka and Binger had called all-in and Binger had won the hand overall but Wasicka's hand had beaten Gold's hand, then Wasicka would not have been eliminated, since he had more chips than Binger: Binger would have ended up with 33 million chips, Wasicka with 14 million chips, and Gold with 42 million chips. If all three players were all-in and Gold had won the hand, of course, the tournament would have been over and Gold would have won because he had more chips than his opponents at the start of the hand. By convention in such situations, the player who started the hand with more chips, which in this case was Wasicka, would get the prize money for second place.

  Suppose all three players went all-in on the flop and that the probability that Gold wins the hand is 17.17%, the probability that Wasicka wins the hand is 53.82%, the probability that Binger wins the hand and Wasicka's hand beats Gold's hand is 24.03%, and the probability that Binger wins the hand and Wasicka and Gold split the side pot is 1.00%. What is the probability that Binger wins the hand and Gold's hand beats Wasicka's hand?

1.2 Suppose that just before the hand described in the beginning of Section 1.1, Gold's probability of winning the tournament was 60/89, and Binger's probability of winning the tournament was 11/89. What is the probability that Wasicka would win the tournament? Which axiom of probability are you using?

1.3 Suppose that events $A_1, A_2, ..., A_n$ are equally likely events and that exactly one of them must occur. Using axiom 3, prove carefully that $P(A_1) = 1/n$.

1.4 Suppose you are at the final table of the WSOP Main Event and you feel that, based on the number of chips you have and your experience, you have a 30% chance of winning the tournament. Is this a Bayesian or frequentist type of probability?

1.5 Find the general addition rule for three events. That is, find a general formula for $P(A \text{ or } B \text{ or } C)$, for any events A, B, and C, in terms of $P(A)$, $P(B)$, $P(C)$, $P(AB)$, $P(AC)$, $P(BC)$, and $P(ABC)$.

1.6 Using the axioms of probability, prove Boole's inequality, which states that $P(A_1 \text{ or } A_2 \text{ or } ... A_n) \leq P(A_1) + P(A_2) + ... + P(A_n)$, for any events $A_1, A_2, ..., A_n$, where n may be a positive integer or ∞.

1.7 In a huge hand from the 2015 WSOP Main Event, with just 14 players left, Justin Schwartz called with 3♠ 3♣, Alex Turyansky raised with Q♥ Q♣, Joe McKeehen called with 6♠ 6♣, and Schwartz called. The flop came 2♦ 3♦ 6♥, giving both McKeehen and Schwartz three of a kind. Schwartz checked, Turyansky bet 700,000, McKeehen called, Schwartz went all-in, Turyansky folded, and McKeehen called. Schwartz would need the 3♥ (and no 6♦) to take the lead. Given these three players' cards and the three cards on the flop, what is the probability of the 3♥ coming on the turn or river? (In the actual hand, the turn and river were 7♣ 5♣. Schwartz was eliminated in 14th place, McKeehen won the huge 22.32 million chip pot and went on to win the tournament and the $7.7 million first place prize.)

1.8 Compare the answer to the previous exercise involving Schwartz, McKeehen, and Turyansky in the 2015 WSOP Main Event to the approximation obtained using the Rule of Four. Why does the Rule of Four yield an underestimate in this case?

CHAPTER 2

Counting Problems

M any probability problems simply involve counting, especially when considering events that are equally likely. For instance, when dealing a single card from a well-shuffled deck, one may ask what the probability is that the card is a heart. If each of the 52 cards is equally likely to appear, then the probability that the card is a heart is simply the number of hearts in the deck divided by 52. Thus the problem boils down to simply counting the number of hearts in the deck and dividing by the number of equally likely possibilities (52). When dealing two cards from a deck, one may ask what the probability is that they are suited connectors (such as 8♠ 7♠). Or, when dealing five cards from a deck, one may wonder what the probability is that they form a flush. These latter two questions may initially appear more complicated than the first question about whether the single card was a heart, but in fact the same simple principle applies to all of these questions: there are a certain number of distinct possibilities for the cards that will be drawn, and each of these possibilities is equally likely, so the desired probability can often be found by simply counting the number of possibilities in question and dividing by the total number of possibilities.

2.1 Sample Spaces with Equally Probable Events

Suppose a sample space consists of a certain finite number n of elements that are all equally likely and that one of the elements must occur. If we want to determine the probability of some event A that contains exactly k of these elements, then the probability of A is simply k/n. Thus, in such cases, computing the probability of an event amounts to *counting*: counting the number of elements in the event in question (k) and counting the total number of elements in the sample space (n).

Example 2.1.1

At the final table of the WSOP Main Event in 2008, Scott Montgomery went all-in against Peter Eastgate, who called. Montgomery had A♦ 3♦ and Eastgate had 6♠ 6♥. The board was A♣ 4♦ Q♠ A♠. At this point, assuming no other information (such as the cards of the other players at the table who folded), what is the probability of Eastgate winning the hand on the river?

Answer Of the 52 cards in the deck, 8 have been exposed. Given no other information, it is reasonable to suppose that the remaining 44 cards are all equally likely to appear on the river. Only two of these 44 cards allow Peter Eastgate to win the hand: the 6♣ or 6♦. Hence, based on this information, the probability of Eastgate winning the hand is 2/44, or 1 in 22. (As it turned out, the river was the 6♦ and Eastgate went on to win the tournament.)

Example 2.1.2

In the 2007 "Paul Featherstone Dream Table" of NBC's *Poker After Dark*, on the final hand, only Gavin Smith and Phil Hellmuth were left. Blinds were $800 and $1600 and Phil Hellmuth was down to just $6700 in chips. When Gavin Smith went all-in, Hellmuth looked at just one of his cards, saw that it was an ace, and called before even looking at the other card. One might ask what the probability of this occurrence would be. That is, given no information about Smith's or Hellmuth's cards, what was the probability of Hellmuth's first card being an ace?

Answer   Each of the 52 cards in the deck is equally likely to be Hellmuth's first card, and exactly four are aces, so the probability of Hellmuth's first card being an ace is simply 4/52 = 1 in 13.

Example 2.1.3

Continuing with the previous example, given that Hellmuth's first card is an ace (and again assuming no information about Smith's cards), what is the probability that Hellmuth has pocket aces?

Answer   Since Hellmuth's first card is an ace, 51 possibilities remain for his other card, and each is equally likely. Exactly 3 of these 51 cards are aces, so the probability that Hellmuth's other card is also an ace is 3/51 = 1 in 17. (By the way, Hellmuth had A7, Smith had 55, and the board came A4975 to end the tournament dramatically.)

Example 2.1.4

To keep opponents guessing, some poker players like to randomize their play, for instance by using the second hands on their watches as suggested by Harrington and Robertie (2004). Suppose that each time you are dealt AJ in first position, you look at your watch and if the second hand is in the northeast quadrant you raise; otherwise you fold. If we assume that the location of the second hand is equally likely to be in any of the four quadrants each time you are dealt AJ in first position, what is the probability that you will raise if you are dealt AJ in first position?

Answer We can consider the sample space for this question to consist of the four quadrants on a watch. By assumption, the second hand is equally likely to be in any of the four quadrants, so we have four equally likely events. Hence the probability in question is simply 1/4.

Example 2.1.5

In a (or perhaps *the*) critical hand from the 2010 WSOP Main Event, on day 8, with only 15 of the original 7319 players left, Jonathan Duhamel, who was at the time just barely the chip leader, raised from 200,000 to 575,000 with J♣ J♥ in the cut-off, and Matt Affleck reraised to 1.55 million with A♣ A♠ on the button. Duhamel reraised to 3.925 million. Affleck called. The flop came 10♦ 9♣ 7♥, Duhamel checked, Affleck bet 5 million and Duhamel called. The turn was the Q♦, Duhamel checked, Affleck went all-in for 11.63 million, and Duhamel faced a tough decision. Based on the board and the players' cards, if Duhamel called, what was his probability of winning the hand?

Answer In order to win, Duhamel needed the river to be a king, jack, or eight. The 44 possible river cards were each equally likely, and there were 4 kings + 2 jacks + 4 eights = 10 cards remaining, so Duhamel's chance of winning is 10/44 or ~22.73%.

In the actual hand, Duhamel called. The river was 8♦, giving Duhamel a straight and awarding him the enormous pot of over 41 million chips. This hand left Duhamel with over 51 million chips, and over a third of the total chips in play (Matthew Jarvis, in second place, had about 29 million chips). Duhamel eventually won the event, becoming world champion and winning the first prize of over $8.9 million.

Example 2.1.6

At the final table of the 2007 WSOP Main Event, with eight players remaining, a seemingly conservative player named Lee Childs was first to act and raised from $240,000 to $720,000. Suppose you know that Childs would only make such a raise with AK, AA, KK, or QQ and that he would be certain to make this raise if dealt any of these hands. Given that he has made this raise, and given no information about his opponents' cards, what is the probability that Childs has AA?

Answer   It is tempting (but incorrect) to reason that only four possibilities exist and hence the probability associated with any one of them must be 1/4. For this reasoning to be valid, however, each of the four possibilities must be *equally likely*. However, these four possible hands occur with different frequencies and hence are not equally likely. A correct solution to this question can be obtained by considering each specific set of two cards, including their suits, in determining the sample space. For instance, one such set would be (A♦, K♠) and another would be (Q♣, Q♥). If we ignore the order of the two cards and consider a set such as (A♦, K♠) to be the same as (K♠, A♦), then these sets are called *combinations*. Now, it is easy to see that each combination of two cards will arise with equal probability, and given the information in this example, our sample space for Childs's hand would contain the following 34 equally likely combinations:

{(A♣, K♣), (A♣, K♦), (A♣, K♥), (A♣, K♠), (A♦, K♣), (A♦, K♦), (A♦, K♥), (A♦, K♠), (A♥, K♣), (A♥, K♦), (A♥, K♥), (A♥, K♠), (A♠, K♣), (A♠, K♦), (A♠, K♥), (A♠, K♠), (A♣, A♦), (A♣, A♥), (A♣, A♠), (A♦, A♥), (A♦, A♠), (A♥, A♠), (K♣, K♦), (K♣, K♥), (K♣, K♠), (K♦, K♥), (K♦, K♠), (K♥, K♠), (Q♣, Q♦), (Q♣, Q♥), (Q♣, Q♠), (Q♦, Q♥), (Q♦, Q♠), (Q♥, Q♠)}.

We can see in this list that 16 of the combinations are AK, 6 are AA, 6 are KK, and 6 are QQ. Since each of these 34 combinations is equally likely, the probability of Childs having AA is 6/34 or about 17.65%. (Childs, incidentally, had Q♥ and Q♠.)

The concepts in the remainder of this chapter can be used to answer the question above more rapidly, without having to list each possible combination of two cards.

A brief word about notation: It is common in the poker literature to use abbreviations such as AK to refer to an ace and a king, or A♦ A♥ to refer to the ace of diamonds and ace of hearts, for example. In contrast, in the probability literature, such two-card combinations are often referred to in parentheses and separated by commas, e.g., (A, J) or (A♦, A♥). We will alternate between these notations in this book, typically using the poker notation for brevity and reserving the parenthetical notation for when it is useful to provide clarity, to emphasize the reference to unordered outcomes, or for enumerating many possible hands, as in the previous example.

2.2 Multiplicative Counting Rule

The following principle may seem obvious and its proof trivial, but the result is very useful for quickly counting the number of ordered outcomes (such as cards dealt from a deck) and is the basis for counting the number of different possible permutations and combinations in general.

The multiplicative counting rule: If there are a_1 distinct possible outcomes for event 1, and for each of them, there are a_2 distinct possible outcomes for event 2, then there are $a_1 \times a_2$ distinct possible ordered outcomes on both. In general, with j events, and with a_i possibilities for event i, the number of distinct ordered outcomes is $a_1 \times a_2 \times \ldots \times a_j$.

Example 2.2.1

To illustrate the meaning of *ordered* outcomes, consider a game in which you and an opponent receive just one card each. Here, the order of the two cards matters, in the sense that the event in which you receive the A♣ and your opponent receives the K♥ is fundamentally different from the event in which you receive the K♥ and your opponent receives the A♣. In shorthand, we might write [A♣, K♥] ≠ [K♥, A♣]. How many different ordered outcomes are possible?

Answer By the multiplicative counting rule, the number of distinct possible ordered outcomes would be 52 × 51, since there are 52 choices for your card, and for each such choice, there are 51 distinct possibilities for your opponent's card.

Example 2.2.2

In the World Poker Tour's $1 million Bay 101 Shooting Star event in 2005, with just four players left, the blinds were 20,000 and 40,000, with antes of 5000. The first player to act was Danny Nguyen, who went all-in for 545,000 with A♦ 7♦. Next to act was Shandor Szentkuti, who called with A♠ K♣. The other two players (Gus Hansen and Jay Martens) folded. The flop was 5♥ K♥ 5♠. Based on the information you have about the flop and these two players' hands only, what is the probability that Nguyen wins the hand? (Note that there is also some chance of a split pot if the turn and river were both 5s or one was a 5 and the other an ace, but this question concerns only the chance that Nguyen will win the entire pot.)

Answer Since these four cards belonging to Nguyen and Szentkuti and three cards on the flop have already been removed from the deck, 45 cards remain for the turn and river. Any of these 45 cards is equally likely to arise on the turn, and no matter which card comes on the turn there are 44 equally probable possibilities for the river. Hence, by the multiplicative counting rule, 45 × 44 = 1980 different possible ordered outcomes exist for the turn and river and all are equally likely. How many of these outcomes allow Nguyen to beat Szentkuti? Nguyen can win only if the turn and river are both 7s. By the multiplicative rule of counting, 3 × 2 = 6 such ordered outcomes are possible: three 7s may come on the turn, and for any of these, exactly two 7s remain as possibilities for the river. Thus the probability of Nguyen winning is 6 ÷ 1980 = 1 in 330. (As it turned out, the turn was the 7♠ and the river was the 7♣, so Nguyen won the hand. Szentkuti was eliminated in fourth place on the next hand; Nguyen went on to win the tournament.)

2.3 Permutations

The examples in the previous chapter involved counting the number of ordered outcomes of two events. In this chapter, we generalize the discussion to ordered outcomes of two or more events. Such ordered outcomes may be called *permutations*—a term most often used to refer to n ordered outcomes of n distinct objects, as in Example 2.3.1. We will use the term generally to mean any k ordered outcomes of n distinct objects.

The multiplicative rule of counting can be used to count the number of distinct permutations that are possible. Let $j!$ refer to the product $j \times (j-1) \times (j-2) \times \ldots \times 1$, with the convention that $0! = 1$. Given n distinct objects, the number of different ordered sets of k distinct choices of the n objects is $n \times (n-1) \times \ldots \times (n-j+1) = n! \div (n-k)!$ for any integers k and n with $0 \leq k \leq n$, since there are n possibilities for the first object chosen, and for each such choice there are $(n-1)$ possibilities for the second object chosen, and so on.

In particular, letting $k = n$, we see that the number of distinct orderings of n distinct objects is

$$n! \div (n - n)! = n!$$

Example 2.3.1

Count the number of different possible ways to order the 52 cards in the deck.

Answer Here $n = 52$ and $k = 52$; based on the formula above, the number of such permutations of the deck is 52!, which is approximately 8×10^{67}.

Note that when choosing k distinct ordered items uniformly at random from a collection of n different choices, each such permutation is equally likely. For instance, when dealing from the $n = 52$ card deck, each ordered collection of k distinct cards is equally likely, so the principle of Section 2.1 can be applied, as in the examples below.

Example 2.3.2

What is the probability that, on a given hand, the five board cards form a straight flush in increasing order? (For example, if the order of the cards is 7♥ 8♥ 9♥ 10♥ J♥ or A♥ 2♥ 3♥ 4♥ 5♥, then we would say they form a straight flush in increasing order, but not if they come J♥ 10♥ 9♥ 8♥ 7♥.)

Answer We first must count the number of distinct orderings of the board that are possible. The number of permutations of five cards among 52 distinct cards is $52!/(52 - 5)! = 311{,}875{,}200$ and each such permutation is equally likely to appear on the board. The number of such permutations that result in a straight flush in increasing order is 40: the first board card must be any A, 2, 3, ..., 10, and for any of these 40 choices for the first board card there is exactly one choice for the remaining board cards such that the board forms a straight flush in increasing order. Thus the probability of such a board occurring is 40 ÷ 311,875,200 = 1 in 7,796,880.

Example 2.3.3

Imagine ranking all 52 cards in some order, such as A♠, A♥, A♦, A♣, K♠, K♥, K♦, K♣, ..., 2♠, 2♥, 2♦, 2♣. What is the probability that, on a given hand, the five board cards come in increasing order of rank?

Answer Note that for any choice of five board cards, the five cards are equally likely to appear in any order. The number of distinct possible ways the five cards can be arranged is 5! = 120 and only one such arrangement will result in the cards appearing in increasing order of rank. Thus the probability the cards will appear in this order is 1/120.

Example 2.3.4

One basic pre-flop rule of thumb is to raise with your best $1/n$ of all possible starting hands if no one has raised yet and n players are left to act behind you (excluding you). Of course, pocket aces is the best starting hand in any position and many of the other best starting hands include an ace. Suppose you are at a 10-handed table where each player uses this strategy. What is the probability that, after the first raiser, the *next* player has A♠ A♥?

Answer First, recall that there are 52! equally likely permutations of the cards in the deck. The problem, then, is to count how many of the permutations result in A♠ A♥ (or A♥ A♠) immediately after the first opening hand. Now imagine temporarily removing the A♠ and A♥ from the deck. There are 50! permutations of the remaining

50 cards. For each such permutation, there is only one place where one could insert the A♠ and A♥, so that after the first player with a raising hand, the next player is dealt A♠ A♥.

For instance, suppose after removing the A♠ and A♥ from the deck, the first 18 cards are (3♣, K♥), (Q♥, 5♦), (Q♦, 7♠), (2♠, 3♦), (10♣, 10♠), (A♦, 4♠), (7♥, 8♦), (8♠, 5♣), and (K♦, J♥). Then one may place A♠ A♥ immediately following the 10♠ to obtain a unique permutation with A♠ A♥ immediately following the first opening hand (10♣, 10♠). Similarly, given a permutation of the original 52-card deck such that the player after the first opening hand has A♠ A♥, one may imagine removing the A♠ and A♥ to obtain a unique arrangement of the remaining 50 cards. The key idea here is that there is a one-to-one correspondence between permutations of the 50 cards (with the A♠ and A♥ removed) and permutations of the full 52-card deck such that the player following the first opening hand has [A♠, A♥], in that order. Thus there are 50! such permutations, and thus an equal number of permutations where [A♥, A♠] follows the first raising hand. So, the probability that the player after the first raiser has A♠ A♥ (or A♠ A♥) is $2 \times 50! \div 52! = 1$ in 1326, which is identical to the probability that the first player is dealt A♠ A♥ (or A♥ A♠).

Note that, in the solution above, the fact that there were 10 players at the table and the fact that many of the opening hands contained aces were irrelevant. Regardless of the number of players and what their opening hands are, the probability of the player after the first raiser having A♠ A♥ is the same as the probability of the first player having A♠ A♥. However, a critical component to the problem is that by assumption someone is sure to raise each hand. If, for example, no one has raised until the small blind, then for the small blind, the number of players left would be $n = 1$ so that the small blind will raise with probability 1/1. Thus, the fact that somebody raised provides essentially no information on whether the next player will have pocket aces. If this is changed, as in Example 2.4.14, then the solution will change slightly as well.

2.4 Combinations

A *combination* is an unordered collection of outcomes. For instance, in Texas Hold'em, how many distinct two-card hands can you be dealt? The answer would be 52×51 if order mattered, i.e., if we considered the case where your first card is A♣ and your second is K♥ distinct from the case where you are dealt these same two cards in the opposite order. However, here it is more natural to consider the order in which the cards arrive as irrelevant. Note that this amounts to counting each possible hand, such as (A♣, K♥), only once, whereas if we were counting permutations we would count each such hand twice. So, since there are 52×51 distinct permutations of two cards, there must be $52 \times 51/2$ distinct combinations of two cards. That is, the number of distinct two-card hands where the order of the two cards does not matter is $52 \times 51/2$.

In general, given n distinct objects, the number of distinct combinations of k of them, i.e., the number of ways to choose k different objects where the order of your choices does not matter, is given by $C(n,k) = n!/[k! \times (n-k)!]$. As seen above, in the case of choosing two distinct cards from a 52-card deck, the number of combinations is $C(52,2) = 52!/[2! \times 50!] = 52 \times 51/2$. $C(n,k)$ is often referred to as "n choose k" and the formula for $C(n,k)$ can be derived quite simply. We have seen that, by the multiplication rule of counting, the number of permutations of k distinct choices out of n different objects is equal to $n!/(n-k)!$, and since there are $k!$ permutations of these k chosen items, in considering permutations we count each such choice exactly $k!$ times. Hence there are $n!/[k! \times (n-k)!]$ unique combinations.

When seeking probabilities involving cards dealt (i.e., drawn without replacement) from an ordinary well-shuffled deck, a useful principle is that each possible combination of cards is equally likely to appear, and each possible permutation of cards is equally likely as well. Thus, many probability problems boil down to simply counting the number of relevant combinations or permutations.

Example 2.4.1

What is the probability of being dealt pocket aces in a hand of Texas Hold'em?

Answer   Each combination of two cards is equally likely, and there are $C(52,2) = 1326$ different combinations of two cards. The number of combinations that involve your having pocket aces is simply $C(4,2) = 6$, since there are four aces and any choice of two of these aces results in your having pocket aces. Thus the probability of being dealt pocket aces is simply 6/1326, or 1 in 221.

Example 2.4.2

What is the probability of being dealt AK in a hand of Texas Hold'em?

Answer   Again, each of the $C(52,2) = 1326$ different two-card combinations is equally likely. How many will give you AK? There are four possible aces you could be dealt; for each such choice, there are four possible choices for the king to go with it. Thus, using the multiplicative counting rule, the number of AK combinations is $4 \times 4 = 16$. Note that this reasoning is valid even though we are considering unordered combinations of the two cards: conceptually, in counting the number of AK combinations, the first event here is not the first card dealt to you. Rather it is the identification of which ace you have; the second event is the identification of the king you have. Each of these events has four possible outcomes, so the probability of being dealt AK is 16/1326, or 1 in 82.875.

Example 2.4.3

Returning to Example 1.5.5, which refers to the 2006 WSOP hand described at the beginning of Chapter 1, what is the probability that Wasicka will make a flush on the turn or the river? Assume Wasicka's cards, his opponents' cards, and the flop are known.

Answer Given the three cards on the flop and the six hole cards belonging to the three players, there are 43 cards remaining that are equally likely to appear on the turn or river. Eight of these remaining 43 cards are spades, and 35 are non-spades. Thus there are $C(43,2)$ equally likely combinations of turn and river cards, and $C(35,2)$ of these combinations include no spades. Thus

P(Wasicka makes a flush)

= P(at least one spade appears on the turn or river)

= $1 - P$ (no spade appears on the turn or river)

= $1 - C(35,2)/C(43,2)$

= $1 - 595/903$

~ $1 - 65.9\% = 34.1\%$

Note that in Example 1.5.5, the Rule of Four was used to yield an approximate answer of 32%. The Rule of Four produces decent approximations for small values of x, but when $x \geq 10$, the approximation rapidly begins to become poor. Figure 2.1 compares the approximation using the Rule of Four (denoted by x in the plot) with the true probability of an out coming on the turn, river, or both (denoted by o), for the case where one knows only one's own cards and the three flop cards.

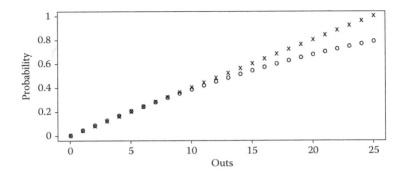

FIGURE 2.1 Approximation (*x*) using the Rule of Four and exact probability (*o*) of an out coming on the turn or river, as a function of the number of outs, assuming knowledge of one's own cards and the three cards on the flop.

Example 2.4.4

In a dramatic hand at the 2008 WSOP Europe Main Event, with 11 players remaining, Chris Elliott and Peter Neff were all-in, the board came 4♣ 3♣ 2♠ 5♣ 6♦ to form a straight on the board, and the two players split the pot. (Neither player had a 7 or two clubs in his hand; Elliott had K♥ Q♥ and Neff had Q♠ Q♣.) One might ask: given no information about the players' cards, what is the probability of the board forming a straight?

Answer For this problem, there is no need to consider the order of the five board cards. There are $C(52,5)$ = 2,598,960 distinct possible combinations of five board cards and all are equally likely to appear. To count how many of these combinations result in a straight, note that there are 10 different possible combinations for the *numbers* appearing on the five cards such that they form a straight (A2345, 23456, 34567, …, 10JQKA). For each such combination, there are four possibilities for the suit of the highest card, and for each such choice there are four possibilities for the suit of the next highest card, and so on, so by the multiplicative counting rule, there are $10 \times 4 \times 4 \times 4 \times 4 \times 4$ = 10,240 distinct possible straight combinations. Hence the probability of a straight appearing on the board is 10,240 ÷ 2,598,960, which is approximately 1 in 253.8, or 0.394%.

Note that in this answer we have included the case where the board forms a straight flush. If we wish to consider the probability that the board will form a straight but not a straight flush, then we must simply subtract the number of straight flush combinations from the numerator of 10,240. There are 10 × 4 = 40 distinct straight flush combinations (10 choices for the numbers on the cards and 4 possibilities for the suit), so the answer would be 10,200 ÷ 2,598,960.

It is very important, when making these types of probability calculations, to keep clear whether you are counting permutations or combinations. This is especially true when computing the probability that some event will occur at least once. It is often helpful in such cases to break the problem down, computing the probability that the event will happen exactly once, and adding to that the probability that the event will happen exactly twice, and so on. Students sometimes get confused in such problems by counting the number of ways that the first such event can occur and multiplying that by the number of ways that the second event can occur. It is important to realize that by ordering events in this manner, one is essentially counting permutations rather than combinations. An example might help clarify the distinction.

Example 2.4.5

Returning to the *Poker After Dark* hand of Example 2.1.2, where Phil Hellmuth needed only to see that one of his cards was an ace in order to call, it seems clear that if *either* of Hellmuth's cards were an ace, he would have called. In light of this one might ask: given no information about the cards of his opponent (Gavin Smith), what is the probability of Hellmuth being dealt at least one ace on this hand?

Answer There are several ways to try to answer this question. One is to note that there are $52 \times 51 = 2652$ distinct possible permutations of two cards that Hellmuth can be dealt, and each is equally likely, so the problem boils down to simply counting the number of such permutations where Hellmuth has at least one ace. Alternatively, one may note that there are $C(52,2) = 1326$ distinct combinations of two cards and Hellmuth is equally likely to be dealt any of these combinations, so again the problem amounts to merely counting the number of combinations involving at least one ace. Since there are half as many combinations as permutations of two cards, it is perhaps a bit easier to work with combinations rather than permutations.

The problem, then, is to count the number of combinations where Hellmuth has at least one ace. Before solving this problem correctly, let us consider a quick but incorrect attempt at a solution. It is easy to come to the wrong conclusion in such problems, reasoning that the number of combinations where Hellmuth has at least one ace is 4×51. The (fallacious) reasoning is as follows: he must have an ace, and there are four choices for this ace; regardless of which ace it was, there are 51 choices left for his other card, and any of these choices will leave him with at least one ace. The problem with this reasoning is that we are essentially double-counting the combinations where Hellmuth has two aces. In this situation we are treating one of his aces as the first card, and the other as the second card, so we are in a sense counting permutations rather than combinations.

To make this explicit, consider enumerating all of the 4×51 outcomes where Hellmuth has at least one ace. We might begin with combinations starting with the ace of clubs: (A♣, 2♣), (A♣, 2♦), etc. At the end of this portion of the list are the combinations (A♣, A♦), (A♣, A♥), and (A♣, A♠). Next we would consider combinations involving the ace of diamonds: (A♦, 2♣), (A♦, 2♦), etc. This portion of the list would contain (A♦, A♣), (A♦, A♥), and (A♦, A♠). It is evident that in the end the entire list would contain (A♣, A♦) as well as (A♦, A♣). However, since we are counting combinations, these outcomes, (A♣, A♦) and (A♦, A♣), are identical and should be counted only once. Each possible pair of aces would be similarly double-counted in this calculation of 4 × 51.

Now let us consider the correct solution to the question of how many combinations of two cards contain at least one ace. As mentioned previously, one helpful trick is to first count the number of outcomes involving exactly one ace and then count the number of outcomes involving two aces. For instance, suppose first that we are counting combinations. The number of combinations of two cards that include exactly one ace is simply 4 × 48 = 192. There are four choices for the ace, and regardless of this choice there are 48 choices for the non-ace. Each pair of choices, such as (A♦, 7♥), for instance, corresponds to one unique combination involving exactly one ace.

Now, how many combinations have exactly two aces? The answer is simply $C(4,2) = 6$, since $C(4,2)$ is the number of combinations of choices of two distinct items out of four different items; the items here are the aces. These six combinations are easy to enumerate: (A♣, A♦), (A♣, A♥), (A♣, A♠), (A♦, A♥), (A♦, A♠), and (A♥, A♠). Thus the total number of combinations of two cards where at least one is an ace is 192 + 6 = 198, and the probability of Hellmuth being dealt at least one ace is 198/1326. Note that the incorrect calculation of combinations given above was 4 × 51 = 204 rather than 198. The incorrect calculation double-counted the six combinations involving two aces.

Note that this problem could equivalently be solved using permutations, rather than combinations. Consider now the order of the cards as they are dealt, and let for example [A♦, 7♥] represent the case that your first card is the A♦ and your second the 7♥. Then by the basic principle of counting, there are 52 × 51 = 2652 such permutations possible, and they are all equally likely. How many contain at least one ace? There are some permutations where your first card is an ace, in which case your second card could be any of the 51 other cards, so there are clearly 4 × 51 = 204 such permutations containing at least one ace. However, there are also permutations where the first card is a non-ace, and the second card is an ace: there are 48 non-aces in the deck, so there are 48 × 4 = 192 such permutations. So, in total there are 204 + 192 = 396 permutations of two cards containing at least one ace. Thus the probability of being dealt at least one ace is 396/2652, which is the same as 198/1326.

For the remainder of this book, we will continue to use brackets, such as [A♦, 7♥], to denote permutations and parentheses, such as (A♦, 7♥), to denote combinations.

Example 2.4.6

In the hand from the 2006 WSOP described in the beginning of Chapter 1, recall that Jamie Gold (with 60 million in chips) called, Paul Wasicka (with 8♠ 7♠ and 18 million in chips) called, Michael Binger (with 11 million in chips) raised to 1.5 million, Gold called, and Wasicka called. The flop was 6♠, 10♣, 5♠. Wasicka checked, Binger bet 3.5 million, Gold went all-in, and Wasicka, facing a difficult decision, decided to fold. Many observers were surprised about Wasicka's decision to fold, since regardless of what cards his opponents had, Wasicka would have had a good chance of winning the hand (among other arguments, some of which we will explore in Section 4.3). From his perspective, Wasicka's worst-case scenario would have been if his opponents had had 9♠ 4♠ and 9♥ 9♦. In this worst-case scenario, what would the probability of Wasicka winning the hand have been?

Answer First, note that since the six cards belonging to the players and three cards on the flop have already been removed from the deck, 43 cards remain for the turn and river. Any combination of 2 of these 43 cards is equally likely to arise on the turn and river and there are $C(43,2) = 903$ such combinations. How many of these combinations enable Wasicka to win the hand? This question involves careful counting. Wasicka would need the turn and river to be (8,8), (7,7), (4,4), (4,x), or (9,y), where x refers to any card that is neither a 4 nor a spade and y refers to any card that is not a 4, 5, 6, 9, 10, or spade. There are $C(3,2) = 3$ combinations in which the turn and river are both 8s {(8♣, 8♦), (8♣, 8♥), (8♦ 8♥)}. Similarly, there are three combinations where the turn and river are (7,7) and three corresponding to (4,4). By the multiplicative counting rule, there are $3 \times 33 = 99$ combinations corresponding to (4,x), since for any of the three choices of 4s, there are 33 choices for x: of the 43 cards remaining for the turn and river, seven are spades and three are 4s, leaving 33 possibilities for x. Similarly, of the 43 cards remaining for the turn and river, seven are spades and twelve others are numbered 4, 5, 6, 9, or 10, leaving $43 - 7 - 12 = 24$ possibilities for y. Hence there are 1×24 combinations for the turn and river corresponding to (9,y). Summing all these terms, we have $3 + 3 + 3 + 99 + 24 = 132$ unordered combinations of turn and river cards yielding Wasicka the winner of the hand, out of 903 equally likely combinations for the turn and river, so the probability of Wasicka winning the hand was $132 \div 903$, or about 14.62%.

Example 2.4.7

Continuing with the previous example, recall that Paul Wasicka had 8♠ 7♠ and the flop was 6♠, 10♣, 5♠. After Wasicka folded, Binger called and showed his cards, which were A♥ 10♥. Gold had 4♠ 3♣. Given this information about the flop and the cards of all three players, if Wasicka and Binger had both called, what is the probability that Wasicka's hand would have beaten both Gold's and Binger's?

Answer There are several cases to consider, and one must be careful not to double-count certain combinations. The table below enumerates the unordered turn–river combinations allowing Wasicka to win. The asterisk is shown in the first row since we wish to count all combinations in which both cards are spades except the combinations (2♠ 3♠) and (A♠ 10♠), in which case Gold would win with a straight flush or Binger would win with a full house, respectively. In the table, a♠ denotes any remaining spade other than the 10♠ or A♠; b represents any remaining non-spade; c is any remaining card other than a 5, 6, 10, ace, or spade; d is any remaining card other than a 10, ace, or spade; $9e$ signifies any 9 other than the 9♠; f is any remaining non-spade other than a 4; and g is any non-spade other than a 4 or 9. Forty-three cards remain as possibilities for the turn and river: 8 are spades and 35 are non-spades. Thus, for instance, when considering the second row of the table, there are $8 - 2 = 6$ possible spades remaining other than the 10♠ and A♠, and for each such choice of a♠ there are 35 possible non-spades to accompany it on the board. Hence, by the multiplicative counting rule, there are $6 \times 35 = 210$ possible combinations associated with this event.

Summing the numbers of combinations on the right in the table below, there are 486 combinations of turn and river cards that would enable Wasicka to win. In total, $C(43,2) = 903$ equally likely turn and river combinations are possible, so the probability of Wasicka winning the hand was $486/903 \sim 53.82\%$.

Unordered Turn–River Outcomes	Number of Corresponding Combinations
♠♠*	C(8,2) − 2 = 26
a♠ b	6 × 35 = 210
10♠ c	1 × 26 = 26
A♠ d	1 × 32 = 32
4 4	C(3,2) = 3
9e, 9e	C(3,2) = 3
4f	3 × 32 = 96
9e g	3 × 29 = 87
8 8	C(3,2) = 3

Incidentally, in the actual hand in question, the turn and river were the 7♣ and Q♠, so Gold won the pot and soon thereafter won the championship and the $12 million first-place prize. Note that if Wasicka had called, he would have made a flush and won this enormous pot.

Example 2.4.8

On one hand from *High Stakes Poker*, Season 2, Gus Hansen had 5♦ 5♣, Daniel Negreanu had 6♠ 6♥, and the board came 9♣ 6♦ 5♥ 5♠ 8♠. Negreanu, before deciding to call Hansen's all-in raise on the river, stated, "Buddy, if I lose this pot, it's a cooler," referring to how incredibly unfortunate he was to have such a powerful hand yet still lose. One question one may ask is: given no information about the board or your opponents' hands, what is the probability of making four of a kind? (To make this question concrete, assume that you never fold and that you see all five board cards every hand.)

Answer A trick for addressing these types of questions is to consider all collections of seven cards (your two cards plus the five on the board) as a single unit, ignoring not only the order of the cards, but also which cards belong to the board and which belong to your hand. Of the $C(52,7) = 133,784,560$ equally likely combinations of these seven cards, how many involve four of a kind? For this part of the problem, we can use the multiplicative rule of counting. There are 13 possibilities for the four of a kind, and for each such choice there are $C(48,3) = 17,296$ possibilities for the other three cards, so there are $13 \times 17,296 = 224,848$ possible seven-card combinations that include four of a kind. Thus the probability of getting four of a kind is $224,848 \div 133,784,560 = 1$ in 595. (Note that in this solution, we are including the case in which the four of a kind is entirely on the board; technically, you still make four of a kind in such a scenario.)

Example 2.4.9

Suppose that you have AK. Given only this information, and assuming that all five board cards will be seen, what is the probability of at least one ace or king coming on the board?

Answer A useful technique for addressing this problem is first to compute the probability of the *complement* of the event in question; by the second axiom of probability, the probability of the desired event is equal to one minus the probability of its complement. The complement of the event in question here is the event that the board contains no ace and no king. There are 50 cards available to come on the board, and thus $C(50,5)$ possible five-card combinations for the board, each equally likely. Forty-four of the 50 cards are neither aces nor kings, so the number of five-card board combinations containing no aces and no kings is simply $C(44,5)$. Thus the probability that the board contains at least one ace or king is

$1 - C(44,5)/C(50,5)$, or approximately 48.74%.

Example 2.4.10

In a hand of heads-up Texas Hold'em, what is the probability that each player will be dealt at least one ace?

Answer This problem is facilitated enormously by using the second axiom of probability.

P(each player has at least one ace)

= 1 − P(player A has no aces *or* player B has no aces)

= 1 − [P(A has no aces) + P(B has no aces)

  − P(A *and* B have no aces)]

= 1 − [$C(48,2)/C(52,2)$ + $C(48,2)/C(52,2)$

  − $C(48,4)/C(52,4)$]

~1.74%, or about 1 in 57.5.

Example 2.4.11

In a very dramatic hand in the 2009 WSOP Main Event, after a flop of 6♣ 6♦ 7♠, Jack Ury bet, Steven Friedlander raised all-in, and Ury called. Friedlander had 7♥ 6♥ for a full house, and Ury revealed 7♦ 7♣ for a higher full house! Suppose you are sure to see the flop on the next hand, no matter what cards you get. What is the probability that you will flop a full house?

Answer As with Example 2.4.8, the key idea in solving such a problem is to consider all combinations of five cards (each representing your two cards plus the three cards on the flop), ignoring which of the five cards represents a card in your hand and which represents a card on the flop. There are $C(52,5)$ = 2,598,960 such combinations of five cards, and each is equally likely.

To count the number of distinct five-card combinations that are full houses, consider any full house, and let events 1, 2, 3, and 4 represent, respectively, the discovery of the number on the triplet, the discovery of the suits of the cards forming the triplet, the discovery of the number on the pair, and the discovery of the suits forming the pair. Then there are 13 possible outcomes for event 1, and for each outcome, there are $C(4,3)$ possible outcomes for the suits on the three cards of the triplet. For each such triplet, there are 12 possible outcomes on the number of the pair, and for each such choice, there are $C(4,2)$ possibilities for the suits on the pair. Hence by the multiplicative counting rule, the number of distinct possible full houses is $13 \times C(4,3) \times 12 \times C(4,2) = 3744$. Thus the probability of flopping a full house is $3744 \div 2{,}598{,}960$, or approximately 1 in 694.17.

Example 2.4.12

(The worst possible beat.) Suppose you have pocket aces, and your opponent has 6♠ 2♠. The first two cards of the flop are revealed and both are aces! At this point, what is your opponent's probability of winning the hand?

Answer Six cards have been removed from the deck (your two aces, your opponent's 6♠ 2♠, and the two aces on the flop), so 46 cards remain. In order for your opponent to win, the last three board cards must be 3♠, 4♠, and 5♠, though not necessarily in that order. Considering *combinations* of the last three board cards, only this *one* combination allows your opponent to win, among the $C(46,3) = 15{,}180$ distinct possible combinations for the last three board cards, of which each is equally likely. Thus P(your opponent wins) $= 1/15{,}180$.

Example 2.4.13

On one hand of *High Stakes Poker*, Season 7, Jason Mercier raised with 9♥ 8♣, Julian Movsesian called with A♣ 9♠, and Bill Perkins called with 5♦ 5♣. When the flop came 2♥ 2♠ 3♠, the host, Norm Macdonald, said "Wow! How often are two 5s an overpair?"

Answer There are 12 cards smaller than 5. Given no other information about your opponents' hands, each of the $C(50,3)$ combinations of three of the remaining cards is equally likely to appear on the flop. So, the probability that the flop contains three cards lower than 5 is simply $C(12,3)/C(50,3)$ ~1.12%, or about 1 in 89.1.

Example 2.4.14

Suppose that, in a given hand of Texas Hold'em with 10 players, at least one of the first nine players is dealt an ace. Consider the player immediately following the first player to be dealt an ace. What is the probability that this next player has A♠ A♥?

Answer Unlike Example 2.3.4, in this problem the information given that one of the first nine players is dealt an ace slightly changes the probability that the next player has A♠ A♥. As in Example 2.3.4, recall that there are 52! equally likely permutations of the cards in the deck. Note that $C(48,18)$ combinations of 18 cards may be dealt to the first nine players so that none of them has an ace. For each such combination, there are 18! possible permutations of the 18 cards dealt and 34! permutations of the other 34 cards not dealt. Thus, there are $52! - C(48,18) \times 18! \times 34!$ permutations of the deck such that at least one of the first nine players has an ace, and each of these permutations is equally likely. Thus, our goal is to count how many of these permutations of the deck result in A♠ A♥ (or A♥ A♠) immediately following the hand of the first player with an ace.

As in Example 2.3.4, we can imagine temporarily removing the A♠ and A♥ from the deck. By the same rationale as in the preceding paragraph, we now have 50! − *C*(48,18) × 18! × 32! permutations of the remaining 50 cards such that at least one of the first nine players dealt has an ace. For each such permutation, there is exactly one place to insert the A♠ and A♥, so that after the first player with an ace, the next player is dealt A♠ A♥, and for any permutation of the original 52-card deck such that someone has an ace and the next player has A♠ A♥, we can imagine removing the A♠ and A♥ in order to obtain a unique arrangement of the remaining 50 cards such that at least one of the first nine players has an ace. Since there is a one-to-one correspondence between permutations of the 50 cards such that at least one of the first nine players has an ace, and permutations of the full 52-card deck such that at least one of the first nine players has an ace and the player following the first hand with an ace is dealt [A♠, A♥], there are 50! − *C*(48,18) × 18! × 32! such permutations. There is an equal number of permutations where [A♥, A♠] follows the first hand with an ace. Thus, the probability that, after the first player with an ace, the next player has A♠ A♥ or A♥ A♠ is 2 × [50! − *C*(48,18) × 18! × 32!] ÷ [52! − *C*(48,18) × 18! × 34!] ~1 in 1846, whereas 1/*C*(52,2) = 1 in 1326. The knowledge that an ace appeared in the first 18 cards makes it slightly less likely that the next player will have pocket aces.

Note that the answer would be simplified considerably if we changed the problem slightly by not indicating that an ace exists in the first 18 cards. For instance, if the dealer simply kept dealing cards from the deck until reaching the first player with an ace, then the probability that the next player would have A♠ A♥ would simply be 2 × 50!/52! = 1/*C*(52,2) = 1 in 1326 by the one-to-one correspondence described above.

Examples 2.3.4 and 2.4.14 are similar to an example in *An Introduction to Probability* by Sheldon Ross in which a dealer reveals one card at a time until the first ace is dealt, and the problem is to determine the probability that the next card is the ace of spades. As in Example 2.3.4, one may consider removing the A♠ and placing it immediately after the first ace in the remaining 51 cards, thus obtaining a unique combination with the A♠ after the first ace corresponding to each of the 51! ways to arrange the other 51 cards, so the probability is 51!/52! = 1/52.

Example 2.4.15

As mentioned in Example 1.5.6, Mark Newhouse notably made the November Nine (the final nine players in the WSOP Main Event) in both 2013 and 2014. Assume that all nine of the finalists from 2013 are among the 6683 entrants in 2014. If all entrants have the same chance to make the November Nine, then what would be the probability of at least one November Nine member from 2013 making it to the November Nine again in 2014?

Answer There are $C(6683,9)$ possible combinations of nine entrants to make the November Nine in 2014, and by assumption each is equally likely to occur. If we omit the nine players who were in the November Nine in 2013, then there are 6674 other players, so there are $C(6674,9)$ possible November Nine combinations in 2014 that do not contain any of the finalists from 2013. Using the second axiom of probability,

P(at least one 2013 finalist making the final nine in 2014)

$$= 1 - P(\text{none of the 2013 final nine making the final nine in 2014})$$

$$= 1 - C(6674,9)/C(6683,9)$$

$$\sim 1.21\%.$$

Example 2.4.16

In a curious hand from *High Stakes Poker*, Season 7, Bill Klein straddled for $1600, and blinds were $400 and $800 with $200 antes from eight players, so $3600 was in the pot before the players looked at their cards. After David Peat and Doyle Brunson called and Vanessa Selbst folded, Barry Greenstein raised to $10,000 with A♣ Q♦. Antonio Esfandiari and Robert Croak folded, Phil Ruffin called, and Bill Klein raised all-in for $137,800 with A♥ K♣. Although the other players folded, Greenstein had only $56,200 left and, somewhat surprisingly, decided to call. Given only the hole cards of Greenstein and Klein, what is the probability that they would split the pot?

Answer This is a complicated problem that may be broken down to relatively simple parts as follows. Given the players' hole cards, there are $C(48,5) = 1,712,304$ possible combinations of boards, each equally likely. We may answer the question by counting how many of these combinations result in splitting the pot. One way to count the number of different boards where they split the pot is first to count the number of ways they can split with the same straight flush, then the number of ways they split with the same four of a kind, and so forth.

Straight flush (29 combinations)—How many different five-card board combinations have Greenstein and Klein splitting the pot with the same straight flush? This requires a straight flush on the board and it must not be improved by their hole cards. This occurs if the board consists of any of the 10 spade straight flushes, plus six club straight flushes (23456 through 78910J), six diamond straight flushes (A2345, 23456, 34567, 45678, 56789, 678910), and seven heart straight flushes. Thus there are $10 + 6 + 6 + 7 = 29$ combinations whereby they split the pot with straight flushes.

Four of a kind (440)—The number of different five-card board combinations whereby they split the pot = $10 \times 44 = 440$. For them to split the pot, the four of a kind must be 2s, 3s, 4s, …, Js, so we have 10 possibilities and for each possible choice, there are 44 possibilities for the other card on the board.

Full house (7969)—There are a few different possibilities that allow the two players to split the pot with the same full house. There can be (1) a full house on the board not involving Qs or Ks; (2) three of a kind on the board and an ace but not three Qs or three Ks; (3) the board could have a pair of aces and another pair lower than Qs and no Qs or Ks; or (4) AAQQx or AAKKy, where x is anything but a Q or K and y is anything but a K. We will count these possibilities individually below.

1. 2200 combinations—The number of board combinations with a full house on the board not involving Qs or Ks = $10 \times C(4,3) \times (9 \times C(4,2) + 1) = 2200$, because for each of the 10 possible numbers on the triplet, there are $C(4,3)$ possible choices of suits for the triplet, and for each choice of triplet there are $9 \times C(4,2)$ possibilities remaining for the pair if the pair is J or lower, plus one possibility where the pair on the board is AA.

2. 3360 combinations—The number of board combinations with three of a kind on the board, but not three Qs or three Ks, and with an ace but not two aces [that outcome was already counted in part (1)] = $10 \times C(4,3) \times 2 \times 42 = 3360$, because there are 10 choices for the number on the triplet, and for each such choice there are $C(4,3)$ choices for the suits on the triplet, and for each of these there are two choices for the ace and 42 choices for the other card on the board.

3. 2160 combinations—The number of board combinations with a pair of aces and another pair lower than Qs, no Qs or Ks, and where

the other pair cannot be three of a kind [that was already counted in part (1)] = $1 \times 10 \times C(4,2) \times 36 = 2160$, because there is only one possible choice of a pair of aces available, and $10 \times C(4,2)$ choices for the number and suits of the other pair. For each such choice, there are 36 possibilities left for the other card on the board so that it is not an A, K, Q, or the number of the pair on the board.

4. 249 combinations—If $x \neq Q$ or K and $y \neq K$, then the number of board combinations of the form AAQQx or AAKKy = $1 \times C(3,2) \times 40 + 1 \times C(3,2) \times 43 = 129$, because there is only one choice for the aces, $C(3,2)$ choices left for the suits on the Q or K, 40 choices for x, and 43 for y.

Thus, the total number of combinations of board cards resulting in a split pot where both players have a full house is $2200 + 3360 + 2160 + 249 = 7969$.

Flush (1277)—These are easy to count. Because the players' lowest cards are Q and K, they cannot make a flush in clubs, diamonds, or hearts where their hole cards do not play, other than the straight flushes described above. Thus, they cannot possibly split the pot with flushes in these suits. We need to count only the number of board combinations in which all five cards are spades but do not form a straight flush because straight flushes were already counted above. The number of these combinations is simply $C(13,5) - 10 = 1277$.

Straight (18,645)—It is easy to count the straight possibilities but quite difficult to count straights that are not flushes for one or both players. Note also that the straight on the board must not be jack or queen high, because otherwise the queen or king from one of their hands would form a higher straight and the two players would not split the pot. The two players split the pot with identical straights as well if the board contains 2345 or KQ J10, provided neither player makes a flush. Because it is easy to double-count outcomes such as 22345, it is convenient to count boards of the type 2345x, where $x \neq 2$, 3, 4, or 5, separately from those of the form 2345x, where $x = 2$, 3, 4, or 5.

We will also break this part of the problem up into several cases: (1) a straight on the board that is not jack or queen high and neither player makes a flush; (2) the board is KQ J10x, where $x \neq$ A, K, Q, J, 10, or 9, and neither player makes a flush; (3) the board is KQ J10y, where $y =$ K, Q, J, or 10, and neither player makes a flush; (4) the board is 2345z, where $z \neq$ A, 2, 3, 4, 5, or 6, and neither player makes a flush; and (5) the board is 2345w, where $w = 2$, 3, 4, or 5, and neither player makes a flush.

1. 6207 combinations—The number of combinations where there is a straight on the board that is not jack or queen high and where neither makes a flush can be enumerated as follows. Consider first the outcome where the board is A2345. There are $2 \times 4 \times 4 \times 4 \times 4$ such boards, and of these, none contain five clubs, two contain exactly four clubs, one contains five diamonds, $1 + C(4,3) \times 3 = 13$ contain exactly four diamonds, none contain five hearts, two contain exactly four hearts, and one contains five spades. There are $1 + C(4,3) \times 3$ containing four diamonds because the 1 corresponds to the particular outcome A♠ 2♦ 3♦ 4♦ 5♦, and $C(4,3) \times 3$ corresponds to the case where the A♦ is on the board and three of the remaining four cards are diamonds, while the other card is a club, heart, or spade. Note that we do not need to worry about subtracting the number of boards containing exactly four spades, because neither player has a fifth spade to make a flush. Thus there are $2 \times 4^4 - 2 - 1 - 13 - 2 - 1 = 493$ combinations of A2345 on the board where neither player makes a flush or straight flush.

   Now consider 23456. There are 4^5 of these combinations and, of these, four are straight flushes,

$C(5,4) \times 3$ contain exactly four clubs, $C(5,4) \times 3$ contain exactly four diamonds, and $C(5,4) \times 3$ contain exactly four hearts. Thus there are $4^5 - 4 - 3 \times C(5,4) \times 3 = 975$ such board combinations and similarly there are 975 each for 34567, 45678, 56789, and 678910.

For 910JQK, there are $4 \times 4 \times 4 \times 3 \times 3$ such board combinations; none contain five clubs, one contains four clubs, none contain five diamonds, one contains four diamonds, one contains five hearts, $C(5,4) \times 3$ contain exactly four hearts, and one contains five spades. Thus, there are $4 \times 4 \times 4 \times 3 \times 3 - 1 - 1 - 1 - C(5,4) \times 3 - 1 = 557$ of these boards to consider.

Finally, for 10JQKA, there are $4 \times 4 \times 3 \times 3 \times 2$ such board combinations; none contain five clubs, none contain four clubs, none contain five diamonds, three contain four diamonds, none contain five hearts, two contain four hearts, and one contains five spades, so there are $4 \times 4 \times 3 \times 3 \times 2 - 3 - 2 - 1 = 282$ of these combinations to consider.

In total, there are $493 + 5 \times 975 + 557 + 282 = 6207$ different board combinations involving a straight on the board, not jack or queen high, where neither player makes a flush or straight flush.

2. 3885 combinations—The number of board combinations involving KQ J10x, where $x \neq$ A, K, Q, J, 10, or 9, is simply $3 \times 3 \times 4 \times 4 \times 28$; none involve five clubs, and 3×7 involve four clubs because there are three non-club kings and seven club possibilities for x. Similarly, none involve five diamonds, 3×7 involve four diamonds, seven involve five hearts, $7 \times (2 + 2 + 3 + 3 + 3)$ involve exactly four hearts, and seven involve five spades. So, the total number of boards involving KQ JTx where neither player makes a flush or straight flush is $3 \times 3 \times 4 \times 4 \times 28 - 3 \times 7 - 3 \times 7 - 7 - 7 \times (2 + 2 + 3 + 3 + 3) - 7 = 3885$.

3. 710 combinations—The number of board combinations of the form KKQ J10, KQQ J10, KQ JJ10, or KQ J1010, where neither of the players makes a flush, can be counted as follows. First, note that with any of these combinations, it is not only impossible to have five of the same suit on the board, but also to have four clubs or four diamonds on the board, because the K♣ and Q♦ are in players' hands. Thus, the only flush possibility is in hearts. There are $C(3,2) \times 3 \times 4 \times 4$ board combinations of the form KKQ J10, and two involve four hearts on the board. Similarly, there are $3 \times C(3,2) \times 4 \times 4$ boards of the form KQQ J10, and two involve four hearts. There are $3 \times 3 \times C(4,2) \times 4$ boards of the form KQ JJ10, and three of these contain four hearts. Similarly, there are $3 \times 3 \times 4 \times C(4,2)$ boards of the form KQ J1010, and three of these contain four hearts. Thus the total is $3 \times C(3,2) \times 4 \times 4 - 2 + 3 \times C(3,2) \times 4 \times 4 - 2 + 3 \times 3 \times C(4,2) \times 4 - 3 + 3 \times 3 \times 4 \times C(4,2) - 3 = 710$.

4. 6343 combinations—The number of board combinations involving 2345z, where $z \neq$ A, 2, 3, 4, 5, or 6, is simply $4^4 \times 26$. Of these, six involve five clubs, and $20 + 6 \times C(4,3) \times 3$ involve four clubs because there are 20 non-club choices for z, and if z is any one of the six club possibilities in the range 7 through Q, then there are $C(4,3) \times 3$ possible choices for the other board cards so that exactly three are clubs. Similarly, six involve five diamonds, $20 + 6 \times C(4,3) \times 3$ involve four diamonds, seven involve five hearts, $19 + 7 \times C(4,3) \times 3$ involve four hearts, and seven involve five spades. So, the total number of boards involving 2345z where neither player makes a flush or straight flush is $4^4 \times 26 - 6 - (20 + 6 \times C(4,3) \times 3) - 6 - (20 + 6 \times C(4,3) \times 3) - 7 - (19 + 7 \times C(4,3) \times 3) - 7 = 6343$.

5. 1500 combinations—The number of board combinations of the form 22345 is simply $C(4,2) \times 4^3$ and, of these, obviously none have five of the same suit, three have four clubs, three have four diamonds,

and three have four hearts. Thus there are $C(4,2) \times 4^3 - 9 = 375$ board combinations of the form 22345 where neither player makes a flush or straight flush. The number of combinations is the same for 23345, 23445, and 233455, so the total is $4 \times 375 = 1500$. So, the total number of board combinations in which the two players split with equivalent straights is $6207 + 3885 + 710 + 6343 + 1500 = 18{,}645$.

Two pairs (51,840)—The number of board combinations involving two pairs on the board, and no queens, kings, or aces on the board, is $C(10,2) \times C(4,2) \times C(4,2) \times 32 = 51{,}840$, because there are $C(10,2)$ choices for the numbers on the two pairs, since they cannot be queens, kings, or aces, and for each such choice, there are $C(4,2)$ possibilities for the suits on the higher pair, and for each such choice, there are $C(4,2)$ possibilities for the suits on the lower pair, and 32 possibilities for the fifth card on the board which may not be the same as either of the two pairs and also may not be a queen, king, or ace.

Note that there are no other possibilities. Because their cards are so high, there is no way the players can split the pot where each player's best five-card hand is three of a kind, one pair, or no pairs. Thus, the total number of combinations of five board cards where the two players split the pot is $29 + 440 + 7969 + 1277 + 18{,}645 + 51{,}840 = 80{,}200$, and $80{,}200/C(48,5)$

$\sim 4.684\%$, or about 1 in 21.35.

In the actual hand, the board came 2♥ 10♠ 7♣ 2♦ 7♠ and the two players split the pot.

Exercises

2.1 In volume 2 of *Harrington on Hold'em* by Dan Harrington and Bill Robertie (2005), the authors argue that sometimes at the end of a tournament, if the blinds are sufficiently high, one must go all-in with any two cards. Suppose that only five players remain in a tournament and that on a certain hand all five players go all-in without even looking at their cards. What is the probability that player A wins the hand?

2.2 An online strategy posted at http://www.freepokerstrategy.com in 2008 advised always going all-in with AK, AQ, AJ, A10, or any pair, and folding all other hands. (The site called this an "Unbeatable Texas Hold'em Strategy.") What is the probability of being dealt one of these hands (AK, AQ, AJ, A10, or any pair)?

2.3 In his book *Play Poker Like the Pros*, Phil Hellmuth (2003) advises beginners to play only with AA, KK, QQ, or AK. What is the probability of being dealt one of these hands?

2.4 What is the probability that you will flop a straight flush on your next hand? (Assume that you never fold; i.e., you are equally likely to see the flop with any two cards.)

2.5 Suppose you have a pocket pair. Given no information about your opponents' cards, what is the probability that you will flop four of a kind? (Assume that you never fold.)

2.6 In the 2014 WSOP Main Event, with just 15 players left, Luis Velador had K♥ 7♥ against Tom Sarra Jr. with 9♠ 9♥, and Velador flopped a flush when the flop came A♥ 4♥ 5♥. Given no information about your cards and assuming you will play any two cards, what is the probability that you will flop a flush (or straight flush) on your next hand? (In the actual hand between Velador and Sarra, the turn was a 7♦ and the river a 3♣. Velador raised before the flop and then bet on the flop, turn, and river and was called all the way to win a 9.09 million chip pot.)

2.7 What is the probability that you will flop an ace high flush on your next hand? (Include the case where the ace is on the board, rather than in your hole cards, and assume that you never fold.)

2.8 Given that both of your hole cards are the same suit, what is the probability that you will eventually make a flush when all five board cards are revealed? Note that technically this includes the possibility that the five board cards are all of the same suit, even if this is not the same suit as your two cards! (Assume that you never fold.)

2.9 What is the probability that you will flop two pairs on your next hand? Note that this includes the case that you have a pocket pair and the flop contains a different pair. (Assume that you never fold.)

2.10 A rainbow flop occurs when the three cards on the flop are all of different suits. Given no information about the players' cards, what is the probability that a given flop will be a rainbow?

2.11 In one hand during Season 4 of *High Stakes Poker*, Jamie Gold had 10♦ 7♠ and Sammy Farha had Q♦ Q♠. After the flop came 9♦ 8♥ 7♣, Gold went all-in and Farha called. At this point, what is the probability that Gold will win the hand, assuming no information about the cards of the other players at the table who have folded? (Note that if the turn and river contain a 6 and a 10, then the pot is split. Do not count this as a win for Gold.)

2.12 What is the probability that on your next hand both your hole cards will be face cards? (A face card is any king, queen, or jack.)

2.13 What is the probability that on your next hand you will be dealt a pocket pair and that both cards will be face cards? (A face card is any king, queen, or jack.)

2.14 What is the probability that on your next hand you will be dealt two face cards but not a pocket pair? (A face card is any king, queen, or jack.)

2.15 Suppose you have A♣ K♣ and are all-in. Given no information about your opponents' cards, what is the probability that you will eventually make a royal flush on this hand, after all five board cards are dealt?

2.16 In a key hand from the 2015 Main Event final table, with just seven players left, Neil Blumenfield raised with 4♥ 4♣, Pierre Neuville called with A♣ K♥, the flop came Q♠ K♠ 4♦, and Blumenfield ended up winning a 14.7 million chip pot. Given that you are dealt a pocket pair, and given no information about your opponents' cards, what is the probability that you will flop either three of a kind or a full house? (Note that technically this includes the case where you have 77 and the flop comes 333, for instance.)

2.17 You are said to have *the nuts* if, given the cards on the board, no other player could possibly currently have a stronger five-card hand than yours. Given only that you have K♣ J♦ and no information about your opponents' cards or the flop, what is the probability that you will flop the nuts?

2.18 What is the worst hand you can have, according to the usual ranking of poker hands and yet still have the nuts on the river? Indicate both the board and your two cards.

2.19 On one hand of *High Stakes Poker* Season 6, after David Benyamine raised to $4200, Phil Galfond reraised to $16,000 as a bluff with K♥ 5♦, and Eli Elezra re-reraised to $40,500 with K♦ K♣. Benyamine folded and Galfond called, surprisingly. The flop was 9♦ 9♣ K♠. At this point, given only the two players' cards and the flop, what is the probability of Galfond winning the hand? What is the probability of the two players splitting the pot? (In the actual hand, after the flop, Elezra bet $33,000 and Galfond called. The turn was the 9♠ and both players checked. The river was the Q♠, Elezra bet $110,000 and Galfond folded his full house!)

2.20 What is the probability of flopping the unbreakable nuts? See Appendix B for a definition and assume you are sure to see the flop.

2.21 As mentioned in Examples 1.5.6 and 2.4.15, Mark Newhouse accomplished the incredible feat of making the November Nine in 2013 and 2014. What is the probability of *all* of the 2013 November Nine winding up in the 2014 November Nine? Assume all nine players participate in both years and that all 6683 participants in 2014 have identical poker strategies.

2.22 Continuing the previous exercise, Newhouse actually wound up in precisely ninth place in both 2013 and 2014. Making the same assumptions as in the previous problem, find the probability that the entire November Nine from 2013 would wind up in exactly the same finishing order in 2014.

2.23 In a key hand from the 2014 WSOP Main Event, with three players left, Jorryt Van Hoof raised to 3.6 million chips with A♦ 5♦, Martin Jacobson reraised to 9.2 million with A♠ 10♣, Van Hoof went all-in for 46.2 million, and Jacobson quickly called. The board came 5♠ 2♥ 10♥ Q♣ Q♠. Jacobson won the 94.6 million chip pot, eliminating Van Hoof, and soon after Jacobson won the event. (a) Given the cards the two players had, what is the probability that a 5 and a 10 would both appear on the flop (more specifically, what is the probability of *at least one* 5 and *at least one* 10 appearing on the flop)? (b) Given the cards the players had and the flop, what was the probability of Van Hoof winning the hand? (Do not include the probability of a split pot.)

2.24 Continuing with Exercise 1.7 involving the hand from the 2015 WSOP Main Event between Justin Schwartz with 3♠ 3♣, Alex Turyansky with Q♥ Q♣, and Joe McKeehen with 6♠ 6♣, where the flop came 2♦ 3♦ 6♥, given these three players' cards and the three cards on the flop, what is the probability of Schwartz winning the hand? (Ignore the possibility of a split pot if the turn and river were a 4 and 5.)

2.25 Continuing Example 1.5.7 from the 2015 WSOP Main Event where Andrew Moreno had A♦ K♦, Dax Greene had A♠ A♥, and the flop came 5♦ K♠ K♥, given the players' cards and the three cards on the flop, what is the probability of Greene winning the hand?

2.26 On day 6 of the 2015 WSOP, Brian Hastings raised on the button with A♦ 9♦, Neil Blumenfield reraised with Q♦ 10♠, and Hastings called. When the flop came 9♣ 2♥ 5♣, Blumenfield bet, Hastings raised, Blumenfield bluffed all-in, and Hastings correctly decided to call. Given both players' hands and the flop, what was the probability of Blumenfield winning the hand? (In the actual hand, the board came K♥ J♠, giving Blumenfield a straight and the 2.7 million chip pot. Hastings was later eliminated in 49th place and Blumenfield ended up getting third place, earning $3.4 million.)

2.27 On day 6 of the 2015 WSOP Main Event, Randy Clinger and David Peters were all-in on a flop of 9♣ 8♦ Q♣. Clinger had A♦ Q♠ and Peters had 10♣ 7♣. (a) Given their hands and the flop, calculate the probability of Peters winning the hand. (b) Using the Rule of Four, estimate the probability of Peters winning given their hands and the flop. (In the actual hand, the turn and river were the 2♠ and 2♦, awarding Clinger the 4.07 million chip pot.)

2.28 On day 6 of the 2015 WSOP Main Event, Joe McKeehen went all-in with A♥ Q♥ and was called by Josh Beckley with A♦ K♣. The board came a dramatic K♦ J♥ 4♦ 6♣ 10♦, giving McKeehen a straight and the 3.84 million chip pot, and McKeehen went on to win the event. Given only McKeehen's cards, what was the probability of him making a straight? (Ignore the possibility of a straight coming on the board, such as when the board is 34567.)

2.29 On day 6 of the 2015 WSOP Main Event, with only 36 players left, Upeshka De Silva went all-in with 9♠ 9♦ and was called by Alex Turyansky with A♦ A♣. When the flop was A♥ 8♥ 4♥, De Silva was in big trouble. Given their cards and the flop, what was (a) De Silva's probability of winning the hand and (b) De Silva's probability of splitting the pot with a flush on the board? (The turn and river were the uneventful 5♠ and 4♣.)

2.30 On day 6 of the 2015 WSOP Main Event, with only 35 players left, after a raise by Alex Turyansky and two calls, Federico Butteroni went all-in from the small blind with 6♥ 6♦, and everyone folded except Chris Brand with 10♥ 10♦. The pot was 3.99 million chips. The flop came 6♠ Q♣ 6♣, giving Butteroni four of a kind. (a) Given only the hands Butteroni and Brand had and the three cards on the flop, what was Brand's probability of winning the hand? (b) Given no information about your cards or anyone else's cards or the flop, what is the probability of flopping four of a kind? (In the actual hand, the turn and river were the uneventful 4♣ and 4♥.)

2.31 During ESPN's broadcast of the 2015 WSOP Main Event final table, when Zvi Stern had J♥ J♣ against Joe McKeehen on the flop of 5♣ Q♠ 5♦, Antonio Esfandiari commented correctly that when you have pocket jacks, at least one overcard will come on the flop about 57% of the time. What about if you have pocket 10s?

2.32 Even when you flop the top two pairs, you can sometimes have very little chance to win. With 117 players left on day 5 of the 2014 WSOP Main Event, after Dan Smith raised to 70,000, Kane Kalas called in the big blind with Q♠ 9♦. The flop came 9♣ Q♣ 8♥, giving Kalas the top two pairs. The only problem was that his opponent, Smith, had Q♥ Q♦. Given the cards of the two players and the flop, at this point what was the probability of Kalas winning the hand? What was the probability of a split pot? (In the actual hand, the turn and river were A♥ and 9♠, and Smith won the 4.21 million chip pot.)

2.33 With 10 players left in the 2014 WSOP Main Event, after Martin Jacobsen tried to raise with A♣ J♣ but put in insufficient chips so it was ruled a call, William Tonking called with J♥ 9♣ in the small blind and Dan Sindelar checked with J♠ 8♠ in the big blind. The flop came 7♣ 8♥ 10♣, giving Tonking a straight. After some betting and raising, Tonking and Jacobsen went all-in and Sindelar folded. At this point, what was the probability of Jacobsen winning the hand? (The turn and river were the 5♦ and 7♦, so Tonking won the 11.25 million chip pot, but Jacobsen ended up winning the event.)

CHAPTER 3

Conditional Probability and Independence

In this chapter, we deal with problems in which some information given may influence the probabilities in question. The resulting conditional probabilities arise frequently in poker problems and in other applications. In the particular case where the observed events do not influence the probabilities of relevant future events, we say the events are independent. This situation rarely occurs in problems involving one particular hand of Texas Hold'em, because the deck is finite and is not reshuffled after each card is dealt, so the appearance of one or more cards substantially changes the distribution for future cards. However, since the deck is reshuffled between hands, what occurs on one hand may typically be considered independent of what occurs on other hands, so problems involving independence can arise when considering collections of hands.

3.1 Conditional Probability

Often one is given information that can provide some insight into the probability that an event will occur, and it can be important to take this information into account. In poker, for instance, you may be able to compute that the probability of a particular player being dealt AA on a given hand is 1/221, but if you were dealt an ace on the same hand, then the probability of your opponent having AA drops substantially, and if you were dealt AA, then the chance of your opponent having AA decreases dramatically. Alternatively, you may note that your probability of being dealt KK is 1/221, but if you have already looked at one card and have seen that it is a king, then your probability of being dealt KK improves considerably. The resulting probabilities are called *conditional*, since they depend on the information received when you look at your cards. We will approach the solutions to the conditional probability problems described above momentarily, but first, let us discuss the notation and terminology for dealing with conditional probabilities.

The conditional probability that event A occurs, given that event B occurs, is written $P(A \mid B)$ and is defined as $P(AB) \div P(B)$. Recall from Chapter 1 that $P(AB)$ is the probability that both event A and event B occur. Thus the conditional probability $P(A \mid B)$ is the probability that both events occur divided by the probability that event B occurs. In the Venn diagram analogy, $P(A \mid B)$ represents the *proportion* of the area covered by shape B that is covered by both A and B. In other words, if you were to throw a pencil at a paper and the pencil were equally likely to hit any spot on the paper, $P(A \mid B)$ represents the probability that the pencil would land in the region covered by shapes A and B, given that it lands in shape B.

If $P(B) = 0$, then $P(A \mid B)$ is undefined, just as division by zero is undefined in arithmetic. This makes sense, since if event B never happens, then it does not make much sense to discuss the frequency with which event A happens given that B also happens.

In some probability problems, both $P(B)$ and $P(AB)$ are given or are easy to derive, and thus the conditional probability $P(A \mid B)$ is also easy to calculate (see Example 3.1.1, which was mentioned at the beginning of this section).

Example 3.1.1

Suppose you have seen that your first card is K♥. What is the conditional probability that you have KK, given this information?

Answer   Let A represent the event that you are dealt KK, and let B be the event that the first card dealt to you is K♥. We seek $P(A \mid B)$, which by definition is equal to $P(AB) \div P(B)$. $P(B)$ is clearly 1/52, since each card is equally likely to appear as the first card dealt to you. Note that in computing $P(B)$, we are ignoring our extra information that our first card is K♥. In general, whenever we use the notation $P(B)$, we mean the probability of B occurring, given *no* special information about this hand. In other words, we may interpret $P(B)$ as the *proportion* of hands in which event B will occur and clearly this is 1/52. $P(AB)$ is the probability that you are dealt one of the permutations [K♥, K♣], [K♥, K♦], or [K♥, K♠], with the cards appearing in the order specified. Thus $P(AB) = 3/(52 \times 51) = 3/2652$, or 1 in 884. The resulting conditional probability $P(A \mid B)$ is therefore $1/884 \div 1/52 = 1/17$.

Like many probability problems, this question can be tackled in different ways. One alternative is to consider that, given that your first card is the K♥, each of the other 51 cards is now equally likely to be dealt to you as your next card. Since exactly three of these will give you KK, the conditional probability of you being dealt KK, given that your first card is the K♥, is simply 3/51 = 1/17.

Example 3.1.2

What is the probability that you are dealt KK, given that your first card is a king?

Answer   The only difference between this problem and Example 3.1.1 is that now your first card can be any king, not necessarily K♥. However, this problem can be handled just as Example 3.1.1, and in fact the answer is the same. Again, let A represent the event that you are dealt KK, and let B be the event that the first card dealt to you is a king. $P(B)$ is $4/52 = 1/13$, and now AB is the event that both cards are kings and the first card is a king, which is the same as the event that both cards are kings. In other words, in this example, if A occurs then B *must* also occur, so $AB = A$. Thus, $P(AB) = P(A) = C(4,2)/C(52,2) = 6/1326$ or $1/221$, and $P(A \mid B) = 1/221 \div 1/13 = 1/17$.

Note that it makes sense that this probability would be the same as in Example 3.1.1. The fact that your first card is K♥ should not make it any more or less likely that you have KK than if your first card were any other king.

Example 3.1.3

What is the probability that you are dealt KK, given that you are dealt a king? That is, imagine you have not seen your cards, and a friend looks at your cards instead. You ask the friend, "Do I have at least one king?" and your friend responds, "yes." Given this information, what is the probability that you have KK?

Answer   Note how the information given to you is a bit different from Example 3.1.2. It should be obvious that the event that your first card is a king is less likely than the event that your first *or second* card is a king. Let A represent the event that you are dealt KK and B the event that at least one of your cards is a king. We seek $P(A \mid B) = P(AB) \div P(B)$ and, as in Example 3.1.2, $AB = A$, so $P(AB) = P(A) = 1/221$.

$P(B) = P$(your first card is a king or your second card is a king)

= P(first card is a king) + P(second card is a king)
   – P(both are kings)

= 4/52 + 4/52 – 1/221 = 33/221.

Therefore, $P(A \mid B)$ = 1/221 ÷ 33/221 = 1/33.

The difference between the answers to Examples 3.1.2 and 3.1.3 sometimes confuses even those well versed in probability. One might think that, given that you have a king, it shouldn't matter whether it is your first card or your second card (just as it did not matter in Example 3.1.1 whether it was a heart) and that therefore the probability of your having KK given that *either* of your cards is a king should also be equivalent, so it is surprising that this is not the case.

The explanation for the difference between Examples 3.1.2 and 3.1.3 can perhaps best be explained in terms of areas corresponding to Venn diagrams (Figure 3.1). Recall that one may equate the conditional probability $P(A \mid B)$ with the probability of a pencil hitting a target A, given that the pencil randomly falls somewhere in shape B. In Examples 3.1.2 and 3.1.3, the size of the target (A) is the same, but the area of shape B is almost

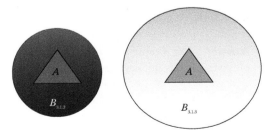

FIGURE 3.1 Venn diagram illustrating probabilities for Examples 3.1.2 and 3.1.3. The triangles (labeled A) represent the event that you are dealt KK. Circles $B_{3.1.2}$ and $B_{3.1.3}$ represent the events that your first card is a king and that you have at least one king, respectively.

twice as large in Example 3.1.3 compared with that in Example 3.1.2.

Note that the three axioms of basic probabilities from Chapter 1 apply to conditional probabilities as well. That is, no matter what events A and B are, $P(A \mid B)$ is always non-negative, $P(A \mid B) + P(A^c \mid B) = 1$, and $P(A_1 \ or \ A_2 \ or \ ... \ or A_n \mid B) = P(A_1 \mid B) + P(A_2 \mid B) + ... + P(A_n \mid B)$ for any mutually exclusive events $A_1, A_2, ..., A_n$. As a result, the rules for basic probabilities given in Chapters 1 and 2 must also apply to conditional probabilities. Thus, for instance, for any events A_1, A_2, and B, $P(A_1 \ or \ A_2 \mid B) = P(A_1 \mid B) + P(A_2 \mid B) - P(A_1 A_2 \mid B)$. Similarly, conditional on B, if events $A_1, ..., A_n$ are equally likely and exactly one of them must occur, then each has conditional probability $1/n$, and if one wishes to determine the conditional probability of some event A that contains exactly k of these elements, then the conditional probability of A given B is simply k/n. Thus, as in Chapter 2, sometimes conditional probabilities can be derived simply by counting.

In the next two examples, we return to the scenarios described earlier in this section.

Example 3.1.4

Suppose you have A♥ 7♦. Given only this information about your two cards, what is the probability that the player on your left has pocket aces?

Answer   Let A represent the event that the player on your left has pocket aces and let B be the event that you have A♥ 7♦. Given B, each combination of two of the remaining 50 cards is equally likely to be dealt to the player on your left. There are $C(50,2) = 1225$ distinct combinations of these 50 cards, so each has conditional probability $1/1225$. The event A contains exactly $C(3,2) = 3$ of these events (A♣ A♦, A♣ A♠, or A♦ A♠), so $P(A \mid B) = 3/1225$ or approximately 1 in 408.33.

Example 3.1.5

In an incredible hand from the final table of the $3000 no-limit Texas Hold'em WSOP event in 2007, with only eight players left, after Brett Richey raised, Beth Shak went all-in with A♥ A♦, and the player on her left, Phil Hellmuth, quickly reraised all-in. Hellmuth had the other two aces. What is the chance of this happening? Specifically, suppose you have A♥ A♦. Given only this information about your two cards, what is the probability that the player on your left has pocket aces?

Answer Again, let A represent the event that the player on your left has pocket aces. Let B = the event that you have A♥ A♦. Given B, each of the $C(50,2)$ = 1225 equally likely combinations of two of the remaining 50 cards has conditional probability 1/1225. The event A now contains just $C(2,2)$ = 1 of these events (namely A♣ A♠), so $P(A \mid B)$ = 1/1225.

Incidentally, it turned out that Richey called with K♣ K♠ and the board came 10♠ 3♦ 7♠ 8♣ 4♣, eliminating Richey in eighth place, and Hellmuth and Shak split the pot.

Example 3.1.6

On day 4 of the 2015 WSOP Main Event, after Mike Cloud raised to 15,000 with A♣ A♠, Phil Hellmuth Jr. called with A♥ K♠, and Daniel Negreanu called from the big blind with 6♦ 4♥. The flop came K♣ 8♥ K♥, giving Hellmuth the lead, and when the hand was over he ended up winning a pot of 161,000 chips. Before the flop, given only the cards Cloud, Hellmuth, and Negreanu had, what was the probability of Hellmuth flopping a better hand than Cloud?

Answer Note first that if the flop came KKA, then Cloud would still be in the lead with a higher full house. The only ways for Hellmuth to have the lead on the flop were for the flop to contain three kings, Q J10, or two kings and x, where x is any card other than an ace or king. Given the six cards the players had, each of the $C(46,3) = 15,180$ possible flop combinations were equally likely. One such combination contains three kings, $4 \times 4 \times 4 = 64$ correspond to Q J10, and $C(3,2) \times 42$ correspond to KKx, so given the hole cards of the three players, the probability of Hellmuth taking the lead on the flop was $(1 + 64 + 3 \times 42) \div 15,180 \sim 1.26\%$.

3.2 Independence

Independence is a key concept in probability, and most probability books discuss enormous numbers of examples and exercises involving independent events. In poker, however, where cards are dealt from a finite deck and hence any information about a particular card tends to change the probabilities for events involving other cards, it is more common for events to be *dependent*. Because the deck is shuffled between hands, one may think of events involving *separate* hands as independent, but for events in the *same* hand, one typically would come to the wrong conclusion by computing probabilities as if the events were independent. It is similarly important in many scientific disciplines to be wary of applying rules for independent events in situations involving dependence.

Independence is defined as follows. Events A and B are *independent* if $P(B \mid A) = P(B)$. Note that, if this is the case, then $P(A|B) = P(A)$ provided $P(B) > 0$, so the order of the two events A and B, i.e., the decision about which event is called A and which is called B, is essentially irrelevant in the definition of independence.

The definition agrees with our notion of independence in the sense that, if A and B are independent, then the occurrence of A does not influence the probability that event B will happen. As a simple example, suppose you play two hands of Texas Hold'em. *A* is the event that you are dealt pocket aces on the first hand, and *B* is the event that you are dealt pocket aces on the second hand. Because the cards are shuffled between hands, knowledge of the occurrence of *A* should not influence the probability of *B*, which is 1/221 regardless of whether *A* occurred or not.

It is generally accepted that the following events may be assumed *independent*:

- Outcomes on different rolls of a die
- Results of different flips of a coin
- Outcomes on different spins of a spinner
- Cards dealt on different poker hands

Sampling from a population is analogous to drawing cards from a deck. In the case of cards, the population is only of size 52, whereas in scientific studies, the population sampled may be enormous. Dealing cards and then replacing them and reshuffling the deck before the next deal is called *sampling with replacement*. When sampling with replacement, events before and after each replacement are independent. However, when sampling *without* replacement, such events are dependent.

When dealing two cards from a deck, for instance, one typically does not record and replace the first card before dealing the second card. In such situations, what happens on the first card provides some information about the possibilities remaining for the second card, so the outcomes on the two cards are *dependent*. If the first card is the ace of spades, then you know for certain that the second card cannot be the ace of spades. Similarly, when sampling at random from a population in a scientific study, one typically samples *without replacement*, and thus technically the outcomes of such samples are dependent. If, for instance, you are measuring people's hand sizes and the first person in your sample has the largest hands in the population, then you know that the second person in your sample does not have the largest hands, so the two hand sizes are dependent. However, when sampling from a large population such as a city, state, or country with several million residents, it is often reasonable to model the outcomes as independent even when sampling without replacement, because information about one observation provides so little information about the next observation. Thus, in scientific studies involving samples without replacement from large populations, technically the observations are dependent, but one typically assumes independence of the observations because calculations performed on the basis of independence are so close to correct. When dealing with a population of 52, however, this is not the case.

A few more general words about independent events are in order. First, if A and B are independent, then so are the pairs (A^c,B), (A,B^c), and (A^c,B^c). This makes sense intuitively: if knowledge about A occurring does not influence the chance of B occurring, then knowledge about A^c should not influence the chance of B either. Mathematically, it is easy to see that for instance $P(A^c \mid B) = P(A^cB)/P(B) = [P(B) - P(AB)]/P(B) = 1 - P(AB)/P(B) = 1 - P(A \mid B)$, which equals $P(A^c)$ if A and B are independent.

3.3 Multiplication Rules

Multiplication rules are useful when calculating the probability that a number of events will all occur, i.e., when considering the probability that event A_1 *and* event A_2 *and* ... *and* event A_k will occur. In such situations, it is important to determine whether the events A_1, A_2, etc., are independent or dependent, although multiplication rules can be used in both cases. In general, for any sets A and B,

$$P(AB) = P(A) \times P(B \mid A),$$

provided $P(A) > 0$ so that $P(B \mid A)$ is defined. This is sometimes called the *general multiplication rule*, and it follows directly from the definition of the conditional probability $P(B \mid A)$. Slightly more generally, for any sets A, B, C, D, ..., as long as all the conditional probabilities are well defined,

$$P(ABCD \ldots) = P(A) \times P(B \mid A) \times P(C \mid AB) \\ \times P(D \mid ABC) \times \ldots.$$

Note that for *independent events A and B*, $P(B \mid A) = P(B)$, so $P(AB) = P(A) \times P(B)$, provided A and B are independent. Thus the general multiplication rule simplifies a bit:

$$P(ABCD \ldots) = P(A) \times P(B) \times P(C) \times P(D) \times \ldots, \\ if \, A, \, B, \, C, \, D, \, \ldots \, are \, independent.$$

We will refer to the above as the *multiplication rule for independent events*.

The next two examples below show the importance of determining whether events are dependent or independent when using the multiplication rules.

Example 3.3.1

Suppose you and I play two hands of poker. Compare the probabilities of these two events:

Event Y: you get pocket aces on the first hand and I get pocket aces on the second hand.
Event Z: you get pocket aces on the first hand and I get pocket aces on the first hand.

Answer Your cards on the first hand and my cards on the second hand are independent, so by the multiplication rule for independent events,

$$P(Y) = P(you \, get \, AA \, on \, the \, first \, hand) \times P(I \, get \\ AA \, on \, the \, second \, hand) \\ = 1/221 \times 1/221 = 1/48{,}841.$$

However, event Z involves your cards and my cards on the same hand, so your cards and my cards are dependent. From Example 3.1.5, we have seen that, given that you have pocket aces on a particular hand, the probability that I also have pocket aces on the same hand is 1/1225. Thus, by the multiplication rule for dependent events,

$$P(Z) = P(\text{you get } AA \text{ on the first hand and I get } AA \text{ on the first hand})$$

$$= P(\text{you get } AA \text{ on the first hand}) \times P(I \text{ get } AA \text{ on the first hand} \mid \text{you get } AA \text{ on the first hand})$$

$$= 1/221 \times 1/C(50,2) = 1/221 \times 1/1225$$

$$= 1/270,725.$$

The difference here is notable. Event *Y* is over 5.5 times more likely than event *Z*.

Example 3.3.2

The following hand was shown on Season 3 of *High Stakes Poker*. After Barry Greenstein, Todd Brunson, and Jennifer Harman folded, Eli Elezra called $600, Sammy Farha (with K♠ J♥) raised to $2600, Shaun Sheikhan folded, Daniel Negreanu called, and Elezra called. With the blinds and antes, the pot was $8800. The flop was 6♠ 10♠ 8♥. Negreanu bet $5000, Elezra raised to $15,000, and Farha folded. Negreanu thought for 2 minutes, and then went all-in for another $88,000. Elezra, who had 8♣ 6♣, called. The pot was $214,800. Negreanu had A♦ 10♥. At this point, the odds on TV showed 73% for Elezra and 25% for Negreanu. (The percentages do not add up to 100% because there is some chance of a split pot. Can you see how?) The two players decided to "run it twice," meaning the dealer would deal the turn and river twice without shuffling and each of the two pots would be worth $107,400. The first turn and river were 2♠ and 4♥, so Elezra won with two pairs. The second time, the turn and river were the dramatic A♥ and 8♦, so again Elezra won, this time with a full house. One might ask what the probability of something like this is. Specifically, given both their hands, the flop, and the first run of (2♠, 4♥), what is *P(Negreanu takes the lead with an A or 10 on the turn and still loses the pot)*?

Answer   First, observe that if a 10 comes on the turn, then Negreanu cannot lose. If an ace comes on the turn, then given this information, there are now 42 available equally likely river cards left and only four (8♦, 8♠, 6♦, and 6♥) would make Negreanu lose the pot. Thus we are seeking

P(A on turn and Negreanu loses)

= *P(A on turn)* × *P(Negreanu loses pot | A on turn)*

= 3/43 × 4/42

= 1 in 150.5.

Note that this is very different from *P(A on turn)* × *P(Negreanu loses)*, which would be about 3/43 × 73% = 5.09% or 1 in 19.6.

Example 3.3.3

Some casinos offer a "jackpot" of several thousand dollars to a player who gets an incredible hand called a *jackpot hand* yet loses the pot. Specifically, according to the definition in one casino in Los Angeles, you have a *jackpot hand* if you meet the following two conditions:

1. You have a full house with three aces, or better. That is, you have a royal flush, straight flush, four of a kind, or a full house with three aces.
2. Your best five-card hand must use both your hole cards.

Given that you make a hand satisfying the first condition, what is the probability that you also satisfy the second? (To make this problem well defined, ignore the possibility of ambiguity as to whether you use your hole cards, such as when you have A5 and the board is AA553, and assume you were all-in before the hand was dealt, so that it is equally likely that you have AQ as 73.)

Answer Imagine taking the seven cards (your two hole cards and the five board cards) and coloring in green the five cards actually used to make your best five-card hand. The question is essentially asking what the chance is that your two hole cards are green. Of the seven cards, five are green, so your first hole card has a 5/7 chance of being green. Given that your first hole card was green, there are now six possibilities left for your second hole card and four of these are green. So,

P(your first hole card is green and your second hole card is green)

= *P(first hole card is green)* × *P(second hole card is green | first hole card is green)*

= 5/7 × 4/6

= 1 in 2.1, or about 47.6%.

Poker players often express chances in terms of *odds* rather than probabilities. For any event A, the *odds of A* = $P(A)/P(A^c)$ and the *odds against A* = $P(A^c)/P(A)$. These ratios are often expressed using the notation $P(A):P(A^c)$ or $P(A^c):P(A)$. For instance, if $P(A) = 1/4$, then $P(A^c) = 3/4$, so the odds of A are 1/4 ÷ 3/4 = 1/3, which would be written as 1:3, and similarly the odds against A would be 3:1. Note that unlike probabilities, odds ratios cannot be multiplied conveniently according to the multiplication rule: if A and B are independent, then the odds against

AB generally do not equal the odds against A times the odds against B. This is one big advantage when working with probabilities instead of odds ratios.

3.4 Bayes' Rule and Structured Hand Analysis

Bayes' rule, which has proven fundamental to many scientific works, originated from Reverend Thomas Bayes' studies of gambling. Bayes' original essay (Bayes and Price 1763) used an example of a man trying to estimate the probability of winning a prize in a lottery, given the results of thousands of trials in the lottery. Bayes' rule is useful when you know what $P(A \mid B)$ is and you seek the reverse conditional probability, $P(B \mid A)$. Specifically, suppose that B_1, B_2, ..., B_n are mutually exclusive events and that exactly one of them must occur. Suppose you want $P(B_1 \mid A)$, but you only know $P(A \mid B_1)$, $P(A \mid B_2)$, etc., and you also know $P(B_1)$, $P(B_2)$, ..., $P(B_n)$.

Bayes' rule states: if B_1, ..., B_n are mutually exclusive events with $P(B_1 \ or \ ... \ or \ B_n) = 1$, then

$$P(B_i \mid A) = P(A \mid B_i) \times P(B_i) \div [\Sigma_j P(A \mid B_j) P(B_j)].$$

Why is this true? Recall from the definition of conditional probability that $P(A \mid B_i) = P(AB_i) \div P(B_i)$. Therefore, $P(AB_i) = P(A \mid B_i) \times P(B_i)$. Also, observe that because exactly one of the events B_i must occur, we may divide the event A into the mutually exclusive events AB_1, AB_2, ..., AB_n (see Figure 3.2). That is, $P(A) = P(AB_1) + P(AB_2) + ... + P(AB_n)$.

So, $P(B_i \mid A) = P(AB_i) \div P(A)$

$$= P(AB_i) \div [P(AB_1) + P(AB_2) + ... + P(AB_n)]$$

$$= P(A \mid B_i)P(B_i) \div [P(A \mid B_1)P(B_1)$$

$$+ P(A \mid B_2)P(B_2) + ... + P(A \mid B_n)P(B_n)].$$

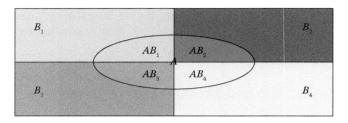

FIGURE 3.2 Venn diagram illustrating Bayes's rule.

Example 3.4.1

A classical application of Bayes' rule has little to do with poker but may be of interest nonetheless and helps clarify the concept. The example concerns tests for rare conditions. Suppose you sample a subject at random from a large population among whom 1% have a certain condition. Suppose you have a test for the condition that is 95% accurate, in the sense that, if a subject actually has the condition, the test will be positive with 95% probability and negative the other 5% of the time, and if the subject does not have the condition, the test will be negative with 95% probability and positive the other 5%. Suppose the randomly selected subject tests positive. Given only this information, what is the probability that the subject in fact has the condition?

Answer   In shorthand, we might write the information given as $P(cond) = 1\%$, $P(+ \mid cond) = 95\%$, and $P(- \mid no\ cond) = 95\%$, where *cond* means the subject really has the condition, the plus sign (+) means the test is positive and the minus sign (−) means the test is negative. We seek $P(cond \mid +)$. Using Bayes' rule,

$$P(cond \mid +) = P(+ \mid cond)P(cond) \div [P(+ \mid cond)$$
$$P(cond) + P(+ \mid no\ cond)\ P(no\ cond)]$$

$$= 95\% \times 1\% \div [95\% \times 1\% + 5\% \times 99\%]$$

$$\sim 16.1\%.$$

This answer is considerably lower than the reader might expect. Considering that the test is 95% accurate, one might suspect that a positive result would yield a 95% chance of having the condition. However, the condition is so rare that the positive result of the test merely increases a patient's chances of having the condition from 1% to 16.1%. To put it another way, since most people do not have the condition and 5% of them will test positive, false positives are very common in this situation and turn out to be more common than true positives in this example. Thus, it is especially desirable for tests of rare conditions to be extremely accurate.

Example 3.4.2

It is widely known among poker players that one should bluff with one's very worst hands, and for some simplified poker scenarios this can be verified mathematically (see Example 6.3.4). On one hand of *High Stakes Poker* Season 3, after an initial raise by David Williams, Brad Booth raised from $1800 to $5800 before the flop, Phil Ivey reraised to $14,000, and Booth called. The flop came 3♦ 7♠ 6♦. Suppose that given what we know up to this point, there is a 2% chance that Booth has a great hand for this flop, such as 45, 33, 66, or 77, and a 20% chance that Booth has a terrible hand. Booth checked, Ivey bet $23,000, and Booth went all-in for $300,000! Suppose you are Phil Ivey, and you believe that if Booth had a great hand, he would make a huge bet like this with 80% probability; if he has a terrible hand, he would make a huge bet like this with 10% probability; and with any other hand, he would not make this bet. Based on all this information and Booth's huge bet, what is the probability that Booth has a great hand?

Answer The assumptions in this hand may be questionable, but with these assumptions it is possible to answer the question directly using

Bayes' rule. Let B_1 = the event that Booth has a great hand, B_2 = the event that Booth has a terrible hand, and A = the event that Booth makes this huge bet on the flop.

$$P(B_1 \mid A) = P(A \mid B_1)P(B_1) \div [P(A \mid B_1)P(B_1) \\ + P(A \mid B_2)P(B_2)]$$

$$= 80\% \times 2\% \div [80\% \times 2\% + 10\% \times 20\%]$$

$$\sim 44.4\%.$$

This problem is similar to Example 3.4.1. The huge bet would seem to indicate a high likelihood that Booth has a great hand, just as the positive test suggests a high likelihood of having the disease. Since great hands are rare, and so is the disease, in each case the ultimate conditional probability is not as large as one might guess. In reality, Booth had 2♠ 4♠ and Ivey folded K♦ K♥!

The computation in Example 3.4.2 is similar to what Harrington and Robertie (2004) refer to as *structured hand analysis*. Given several possible hands for an opponent, you may assign probabilities (or weights) to each of those hands based on their perceived likelihood in your opinion and/or experience. You may then proceed to use Bayes' rule to calculate your chance of winning the pot in a showdown, as in Example 3.4.3.

Example 3.4.3

Phil Gordon (2006, pp. 189–195) discusses a hand from the 2001 WSOP Main Event, with only 13 of the original 613 players remaining. The table in question had six players, and blinds were $3000 and $6000 with $1000 in antes from each player. Mike Matusow raised to $20,000 from under the gun, Phil Gordon reraised to $100,000 with K♣ K♠ in the cut-off seat, and Phil Hellmuth re-reraised all-in to $600,000 from the small blind. Matusow folded, and Gordon faced a tough decision. Suppose Hellmuth would make this re-reraise 100% of the time in this situation if he had AA, but would only make it with 50% probability if he had KK and only with 10% probability if he had AK or QQ. With all other hands, suppose that Hellmuth would not have re-reraised in this scenario. Given Gordon's cards and the fact that Hellmuth made the re-reraise (and ignoring the cards of Matusow and the other players who folded), find (1) the distribution of hands Hellmuth may have and their corresponding probabilities, and (2) the probability of Gordon winning the hand if he calls, based on the answer to (1) and the poker odds calculator at cardplayer. com. (In the actual hand, Gordon folded his K♣ K♠, and Hellmuth showed his pocket aces before collecting the pot.)

Answer

1. We will first determine the distribution of Hellmuth's hands, given our available information. Let B_1 denote the event that Hellmuth has pocket aces; B_2 be the event that Hellmuth has KK; B_3 be the event that Hellmuth has AK; and B_4 be the event that Hellmuth has QQ. Given only knowledge of Gordon's cards, $P(B_1) = C(4,2)/C(50,2) = 6/1225$, $P(B_2) = 1/1225$, $P(B_3) = (4 \times 2)/C(50,2) = 8/1225$, and $P(B_4) = C(4,2)/C(50,2) = 6/1225$. Let A represent the event that Hellmuth makes the re-reraise. We are given that $P(A \mid B_1) = 100\%$, $P(A \mid B_2) = 50\%$, and $P(A \mid B_3) = P(A \mid B_4) = 10\%$. Using Bayes' rule, we may calculate:

$$P(B_1 \mid A) = P(A \mid B_1)P(B_1) \div [P(A \mid B_1)P(B_1) \\ + P(A \mid B_2)P(B_2) + P(A \mid B_3)P(B_3) \\ + P(A \mid B_4)P(B_4)]$$

$$= 100\%(6/1225) \div [100\%(6/1225)$$
$$+ 50\%(1/1225) + 10\%(8/1225)$$
$$+ 10\%(6/1225)]$$

$$= 6/1225 \div [7.9/1225]$$

$$= 6/7.9 \sim 75.95\%.$$

Similarly,

$$P(B_2 \mid A) = 50\%(1/1225) \div [100\%(6/1225)$$
$$+ 50\%(1/1225) + 10\%(8/1225)$$
$$+ 10\%(6/1225)]$$

$$= 0.5/1225 \div [7.9/1225]$$

$$= 0.5/7.9 \sim 6.33\%.$$

$$P(B_3 \mid A) = 10\%(8/1225) \div [100\% (6/1225)$$
$$+ 50\%(1/1225) + 10\% (8/1225)$$
$$+ 10\% (6/1225)]$$

$$= 0.8/1225 \div [7.9/1225]$$

$$= 0.8/7.9 \sim 10.13\%.$$

$$P(B_4 \mid A) = 10\% (6/1225) \div [100\%(6/1225)$$
$$+ 50\% (1/1225) + 10\%(8/1225)$$
$$+ 10\% (6/1225)]$$

$$= 0.6/1225 \div [7.9/1225]$$

$$= 0.6/7.9 \sim 7.59\%.$$

2. Using the poker odds calculator at card-player.com and noting that technically the question asks about Gordon's probability of *winning* (not tying) the hand in a showdown, the probability of Gordon's K♣ K♠ beating AA is approximately 17.82%, the probability of K♣ K♠ beating K♦ K♥ is approximately 2.17%, the probability of K♣ K♠ beating

AK is approximately 68.46%, and K♣ K♠ beats QQ with approximately 81.70% probability. (The suits of the opponents' cards matter somewhat, and the preceding sentence involves averaging over the different possibilities.) Let C = the event Gordon wins the hand if he calls. Given event A, the probability of C is: $P(C$ and $B_1 \mid A) + P(C$ and $B_2 \mid A) + P(C$ and $B_3 \mid A) + P(C$ and $B_4 \mid A)$

$$= P(C \mid A, B_1) P(B_1 \mid A) + P(C \mid A, B_2) P(B_2 \mid A) + P(C \mid A, B_3) \, P(B_3 \mid A) + P(C \mid A, B_4) P(B_4 \mid A)$$

$$\sim 17.82\% \times 75.95\% + 2.17\% \times 6.33\% + 68.46\% \times 10.13\% + 81.70\% \times 7.59\%$$

$$\sim 26.81\%.$$

Exercises

3.1 Suppose some event A has probability p. (a) Let O_A = the odds of A. Find a general equation for p in terms of O_A. (b) Suppose the odds of A are 1/10. What is p? (c) Suppose the odds against A are 5:1. What is p?

3.2 Suppose A and B are independent events. Let $O_{A'}$ = the odds against A and $O_{B'}$ = the odds against B. What are the odds against AB? Express your answer (a) in terms of $P(A)$ and $P(B)$, and (b) in terms of $O_{A'}$ and $O_{B'}$.

3.3 What is the probability that both your hole cards are clubs, given that both of your hole cards are black?

3.4 Show that for any event B and any events $A_1, A_2, \ldots$ the conditional probabilities $P(A_i \mid B)$ satisfy the axioms of probability.

3.5 Let A be the event that both of your two hole cards are aces and let B be the event that your two hole cards are black. Are A and B independent?

3.6 Let A be the event that both your two hole cards are face cards and let B be the event that your two hole cards form a pair. Are A and B independent?

3.7 In a critical hand from the 2008 WSOP Main Event, Dennis Phillips called the $300,000 chip big blind and Ivan Demidov raised to $1.025 million with A♣ Q♣. Phillips reraised to $3.525 million, and Demidov re-reraised to $8.225 million in chips. This was a risky play by Demidov. Suppose Phillips would limp in and then reraise with AA, KK, QQ, or AK. Given that Phillips limped and then reraised and that Demidov has A♣ Q♣, (a) what is the probability that Phillips has AA? (b) Given the range of possible hands for Phillips and their probabilities, if Phillips and Demidov go all-in before the flop, what is the probability that Demidov will win the hand? (Use an online poker odds calculator such as www.cardplayer. com/poker_odds/texas_holdem to help you find, for instance, the probability of K♣ K♦ beating A♣ Q♣.)

3.8 Continuing with the previous exercise, suppose Phillips would always limp and reraise with AA or KK and would only reraise with 50% probability if he had QQ or AK. Now, given that Phillips limped and reraised, what is the probability that he has AA? If they go all-in before the flop, what is the probability that Demidov will win the hand? (Incidentally, in the actual hand, Phillips had A♥ K♣ and called. The flop came 8♦ 10♣ J♠, Phillips bet $4.5 million, Demidov raised all-in for $13.38 million, and Phillips folded.)

3.9 Suppose your opponent looks at her cards and then looks at her chips immediately. Caro (2003) describes this as a *tell*, indicating she has a very strong hand. Suppose you know that she would display a tell with 100% probability if she had AA or KK, with 50% probability if she had AK, and never with any other hand. Given only this information, what is the probability that she has AA?

3.10 Prove that if two events A and B are independent, then so are A^c and B^c.

3.11 Think of an example of three events A, B, and C, such that A and B are independent, and A and C are independent, yet A and BC are not independent.

3.12 Consider fixing two of the 2013 November Nine, such as for instance ninth-place finisher Mark Newhouse and 2013 Main Event champion Ryan Reiss. If we assume both players are among the 6683 participants in 2014 and that all players use the same poker strategy, what is the probability that *both* Newhouse and Reiss would make the 2014 November Nine?

3.13 In an interesting hand on day 4 of the 2015 WSOP Main Event, Jae Kim called 6000 from early position with K♦ J♦, Lance Harris raised to 18,000 from late position with 9♥ 9♣, and Kim called. The flop was 10♦ 5♦ 6♦, so Kim flopped a flush. What is the probability of flopping a flush, given that you have two cards of the same suit? Compare with the probability of flopping a flush given no knowledge of your cards (Exercise 2.6). Assume no knowledge of your opponents' cards.

3.14 What is the probability of being dealt AK, given that both of your hole cards are red? Compare with the probability of being dealt AK given no extra information.

3.15 What is the probability of being dealt AA, given that both of your hole cards are red? Compare with the probability of being dealt AA given no extra information.

3.16 On day 6 of the 2015 WSOP, with 35 players remaining, after a raise by Alex Turyansky and two calls, Federico Buttoroni went all-in from the SB with 6♥ 6♦ and everyone folded except Chris Brand with 10♥ 10♦. The pot was 3,990,000 chips. The flop came 6♠ Q♣ 6♣, giving Buttoroni four of a kind. Given both players' cards and the three board cards, what is the probability of Brand winning the hand? (He didn't; the turn and river were 4♣ and 4♥.)

3.17 Continuing the previous problem, given only Buttoroni's and Brand's cards, what is the probability of Buttoroni flopping four of a kind?

3.18 Suppose when a certain player is under the gun, she raises whenever she has any pair and raises 50% of the time when she has AK or AQ. With any other two cards, she raises under the gun 10% of the time. Given that she has raised under the gun, what is the probability that she has a pair?

3.19 Under the assumptions of the previous exercise, given that she has raised under the gun, what is the probability that she has AK?

3.20 On day 5 of the 2015 WSOP Main Event, Justin Bonomo raised to $55,000 with 5♠ 5♦, Matt Waxman called with A♠ K♠, and the big blind, Antonio Payne, called with A♣ 4♣. The pot was 213,000 chips. The flop was 5♥ K♥ 3♥, giving Bonomo three of a kind and Waxman top pair. Given only the cards Waxman and Bonomo had, calculate the probability of the flop giving Bonomo three of a kind and his opponent Waxman top pair.

3.21 Continuing with the hand from the previous exercise, on the flop, Bonomo bet 125,000, Waxman called, and Payne folded. The turn was 2♣, so Payne would have made a straight and the river was Q♠, so Payne would have won with a straight. However, given just the three players' cards and the three flop cards and assuming none of the players would fold, (a) what was the probability of Payne winning the hand in a showdown? (b) Under the same conditions, what is the probability of Waxman and Payne splitting the pot?

3.22 Late in the 2015 Main Event, after Marcus van Opzeeland raised to 53,000, Salvatore DiCarlo reraised to 132,000 with A♣ K♥, Daniel Negreanu re-reraised to 332,000 with K♠ K♣, and van Opzeeland folded. DiCarlo then re-re-reraised all-in to 1,229,000, and Negreanu called. The flop came 10♦ 3♦ 8♦, raising the possibility of a split pot if the turn and river were both diamonds as well. Given only the cards DiCarlo and Negreanu had and the three cards on the flop, what is the probability of a split pot? (In the actual hand, the turn was 4♦, creating additional suspense, and the river was 7♣, allowing Negreanu to win the roughly 2.5 million chip pot.)

3.23 On day 5 of the 2015 WSOP Main Event, after a raise from Daniel Negreanu, Chad Power called with 10♦ 10♣, Feder Holz reraised with J♠ 8♦, Negreanu folded, and Power called. The flop came 3♥ 10♠ 3♦, giving Power a full house. Holz then continued with two more large bluffs, and Power won the 1.655 million chip pot. (a) Given only the players' cards, what is the probability that Power would flop a full house? (b) Given both players' cards and the three flop cards, and assuming neither player would fold, what was the probability of Holz winning the hand?

3.24 With about 130 players left on day 5 of the 2014 WSOP Main Event, there was a hand where, after the first few players folded, there were six players remaining and five of them had pocket pairs: Brian Roberts had J♥ J♣, Greg Himmelbrand had 6♠ 6♥, Robert Park had 4♥ 4♣, Adam Lamphere had 10♠ 10♥, and Jack Schanbacher had 10♦ 10♣. On any particular hand, let A = the event that you have a pocket pair and let B = the event that the player on your right has a pocket pair. (a) Compute the conditional probability, $P(A \mid B)$. (b) Compare with the unconditional probability, $P(A)$. Which is bigger? (c) Are A and B independent? (This problem is continued in Exercise 7.16.)

3.25 With just 15 players left in the 2015 WSOP Main Event, Alex Turyansky raised with Q♥ 9♥, Daniel Negreanu called from the small blind with A♣ J♣, and Josh Beckley called from the big blind with 10♥ 8♥. The flop came an incredible J♥ Q♦ 9♦, giving Turyansky two pairs and Beckley a straight. Beckley and Turyansky were quickly all-in, and when the turn and river came 6♦ K♥, Beckley took the 13.37 million chip pot. Given only that you have 10♥ 8♥, what is the probability that you will flop a straight (or straight flush)?

3.26 With only 10 players left in the 2015 WSOP Main Event, after Josh Beckley went all-in with A♠ Q♥, Tom Cannuli folded 9♦ 9♠, and Joe McKeehen called with 9♥ 9♣. After the flop of 2♦ A♣ 6♣, McKeehen was suddenly way behind. Given the flop and the six cards belonging to Beckley, what was the probability of McKeehen winning? (The turn and river were the uneventful 4♥ and 4♦, so Beckley won the 12.85 million chip pot.)

3.27 When you flop a pair and a flush draw and are up against just one opponent, you have a decent chance of winning no matter what your opponent has. For example, with just five players left in the 2015 WSOP Main Event, after Neil Blumenfield raised to 2 million with A♣ 4♣, everyone folded to Joe McKeehen, who called with 5♣ 3♣. The flop was 3♥ Q♣ J♣, giving McKeehen a pair and a flush draw against Blumenfield's higher flush draw. (a) Given the two players' cards and the flop, what was McKeehen's probability of winning the hand? (b) Given only that you have 5♣ 3♣ and no other information, what is the probability that you will flop a pair and a flush draw? (In the actual hand, Blumenfield bet 2.5 million on the flop, McKeehen called, and the turn and river were the 5♠ and 3♠, giving McKeehen a full house.)

3.28 What is the P(on your next hand you will be dealt K♥ J♣ and flop a straight)?

3.29 What is P(on your next hand you will be dealt 5♣ 3♣ and flop the nuts)?

CHAPTER 4

Expected Value and Variance

There are few concepts more fundamental either to poker or to science than expected value and variance, which are topics taken up in the context of discrete random variables in Sections 4.2 and 4.5. Before introducing these subjects, we begin with preliminary definitions and simple examples of discrete random variables.

4.1 Cumulative Distribution Function and Probability Mass Function

A *variable* is an item that can take different numeric values. A *random variable* can take different numeric values with different probabilities. We say the random variable X is *discrete* if all of its possible values can be listed. If instead X can take any value in an interval such as $[0,1]$, then we say X is *continuous*. The *distribution* of X means all the information about the possible values X can take, along with their probabilities. Any random variable has a *cumulative distribution function* (cdf): $F(b) = P(X \leq b)$. It follows from the axioms of probability that for any cdf F,

$$\lim_{b \to -\infty} F(b) = 0 \text{ and } \lim_{b \to \infty} F(b) = 1.$$

If X is discrete, then information on its distribution can be summarized using a *probability mass function* (pmf): $f(b) = P(X = b)$. Note that by the second axiom of probability, for any pmf f, $\Sigma f(b)$ must equal 1.

Example 4.1.1

Suppose you play one hand of Texas Hold'em. Let $X = 1$ if you are dealt a pocket pair and $X = 0$ otherwise. What is the pmf of X? What is the cdf of X?

Answer First, note that the number of two-card combinations that give you a pocket pair is $13 \times C(4,2)$, because there are 13 possibilities for the pair's number and for each such choice there are $C(4,2)$ possible combinations for the suits on the two cards. Thus the probability of being dealt a pocket pair is $13 \times C(4,2) \div C(52,2) = 1/17$.

So, for the pmf, $f(1) = 1/17$, $f(0) = 16/17$, and $f(b) = 0$ for all other b. For the cdf,

$$F(b) = 0 \text{ for } b < 0,$$

$$F(b) = 16/17 \text{ for } 0 \le b < 1,$$

$$F(b) = 1 \text{ for } b \ge 1.$$

Such random variables that can only take the values 0 or 1 are called *Bernoulli random variables* and are discussed further in Section 5.1.

Example 4.1.2

Suppose you are one of three players remaining in a tournament. The payouts are $1000 for first place, $500 for second place, and $300 for third place. You feel that you have a 20% chance to end up in first place, a 35% chance for second place, and a 45% chance for third place. Let X be your payout in dollars. What is the pmf of X? What is the cdf of X?

Answer The pmf of X is $f(300) = 45\%$, $f(500) = 35\%$, $f(1000) = 20\%$, and $f(b) = 0$ for all other b. The cdf of X is:

$F(b) = 0$ for $b < 300$.

$F(b) = 45\%$ for $300 \le b < 500$.

$F(b) = 80\%$ for $500 \le b < 1000$.

$F(b) = 100\%$ for $b \ge 1000$.

4.2 Expected Value

Expected value is not only fundamental to the analysis of real, scientific data, but also perhaps the most important principle in the analysis of poker. There are several reasons why expected value applies to poker in particular.

- Tournaments—Some results from game theory suggest that in symmetric winner-take-all tournament competitions, the optimal strategy uses the *myopic rule*: given any choice of options in the game, one should always choose the one that maximizes the *expected value* of the number of chips one has.
- Law of large numbers—Probability theory indicates that if you repeat *independent and identically distributed (iid)* trials over and over, your long-term average will ultimately converge to the *expected value*. Thus, if you play the same game repeatedly, it may make sense to try to maximize the *expected value* of your winnings. This subject is covered further in Section 7.3.
- Checking results—A good way to check whether someone is a long-term winning or losing player, or to verify if a given strategy works or not, is to check whether the sample mean is positive and to see whether it has converged to the *expected value*. This topic is discussed further in Section 7.5.

For a discrete random variable X with pmf $f(b)$, the *expected value* of X is $\Sigma\ b\ f(b)$, where the sum is over all possible values of b. The expected value is also sometimes called the *mean* and is denoted by $E(X)$ or μ. As with most probability texts, we use the notations $E(X)$, $E\{X\}$, and $E[X]$ interchangeably. The only reason to favor the use of one particular set of brackets over another is for clarity; no difference in meaning is intended.

Example 4.2.1

For the random variable X defined in Example 4.1.1, what is $E(X)$?

Answer $f(0) = 16/17$ and $f(1) = 1/17$, so $E(X) = (0 \times 16/17) + (1 \times 1/17) = 1/17$.

In general, for any Bernoulli random variable X, $E(X) = p$, as in Example 4.2.1.

Example 4.2.2

For the random variable X defined in Example 4.1.2, what is $E(X)$?

Answer $f(\$300) = 45\%, f(\$500) = 35\%, f(\$1000) = 20\%$, so $E(X) = (\$300 \times 45\%) + (\$500 \times 35\%) + (\$1000 \times 20\%) = \$510$.

Note that for any random variable X and any constants a and b, if $Y = aX + b$, then $E(Y) = \Sigma(ak + b)\ f(k) = a\Sigma kf(k) + b\Sigma f(k) = aE(X) + b$, since $\Sigma f(k) = 1$ because f is a probability mass function. Thus, for instance, multiplying the potential profits and losses in a game by three amounts to tripling one's expected value, though as we will see in Section 4.5 it means tripling the standard deviation as well.

The expected value of X represents a sort of *best guess* at the value it will take. Note, however, that the expected value might be a number that X cannot possibly take, as in

Examples 4.2.1 and 4.2.2. In the latter example, X cannot be $510: it must be $300, $500, or $1000. Nevertheless, $510 represents a *weighted average* of the values X can take, weighted by their probabilities. $E(X)$ can also be interpreted as the *long-term average* of the values X will take if independent trials are observed indefinitely. When X represents a measure of value or worth, $E(X)$ may be interpreted as a *fair price* to pay for the opportunity to obtain X in return, in the sense that if this price is paid per independent trial, then one's average profit per trial will converge to zero. The following example may illustrate this further.

Example 4.2.3

In 1971, Johnny Moss won the WSOP, which was a single winner-take-all tournament with only seven participants. Suppose you enter such a tournament with a total of seven players, and suppose all are of equal ability and each player has an equal chance of winning. If the player who wins receives a total of $100,000 and if X = your winnings in the tournament, what is the expected value of X?

Answer X will be 0 if you lose and $100,000 if you win, so $E(X) = (\$100,000 \times 1/7) + (\$0 \times 6/7) \sim$ $14,285.71.

Note that the value of X will be 0 or $100,000 in the end. Although the *expected* value might be $14,285.71, in no particular case will the value of X ever be close to $14,285.71. Nevertheless, $14,285.71 would be a fair price for your entry into the tournament in the sense that, if you pay $14,285.71 to enter and play such tournaments indefinitely (and their outcomes are independent of each other, which seems reasonable enough), then your average profit per tournament will ultimately converge to zero.

While the entry fee in 1971 was only $5000, note that the winnings, even including endorsements and indirect earnings, were most likely much less than $100,000. Incidentally, although the terms *fair price* and *fair game* are often used by probabilists to mean games in which the mean profit for the players is zero, you will be hard pressed to find a casino that would agree with this definition of fairness! An important concept we will return to in Section 4.5 is $E(X^2)$. By this expression, we mean the expected value of Y, where regardless of what X is, Y is always equal to X^2. Note that $E(X^2)$ generally does not equal $[E(X)]^2$. The next example may help to clarify this point.

Example 4.2.4

Suppose that, on your next hand of Texas Hold'em, if you are dealt a pocket pair, the casino will pay you $10 and will pay your friend $100. If you are not dealt a pocket pair, you and your friend both get $0. Let X be your profit and Y the profit for your friend. What are the expected values of X and Y? What is $E(X^2)$?

Answer The probability of your getting a pocket pair is 1/17 (see Example 4.1.1). Thus

$E(X) = (\$10 \times 1/17) + (\$0 \times 16/17) = \$10/17 \sim \$0.588$.

$E(Y) = (\$100 \times 1/17) + (\$0 \times 16/17) = \$100/17 \sim \$5.88$.

Note that $Y = X^2$: if $X = \$0$ then $Y = \$0$, and if $X = \$10$ then $Y = \$100$, so Y always equals X^2. Thus $E(X^2) = E(Y) \sim \$5.88$.

The purpose of the previous example is merely to clarify what we mean by X^2 and $E(X^2)$. Note that in this example $E(X^2) \sim \$5.88$ and $[E(X)]^2 \sim \$0.588^2 \sim \$0.346$, so $E(X^2) \neq [E(X)]^2$. In fact, $E(X^2) = [E(X)]^2$ only if X is a constant, i.e., if there is some value c where X equals c with 100% probability.

Example 4.2.5

Near the end of some tournaments, players occasionally decide to make a deal and split the prize money rather than play to the end. Usually the players choose to divide the remaining prize money proportionately according to how many chips they have. This would be fair in terms of expected value for a symmetric winner-take-all tournament, but it usually disproportionately benefits the chip leader in standard non-winner-take-all tournaments. For instance, consider the situation immediately prior to the 2006 WSOP hand described at the beginning of Chapter 1. Suppose that at this point Jamie Gold, who had 60 million of the 89 million in chips, had a 67.4% chance of winning the $12 million first-place prize, a 25.0% chance of winding up in second place and winning $6.1 million, and a 7.6% chance of ending in third place and winning $4.1 million. Compare Gold's *expected* winnings with or without a proportional chip deal.

Answer   Without a chip deal, Gold's expected winnings are ($12 million × 67.4%) + ($6.1 million × 25.0%) + ($4.1 million × 7.6%) ~ $9.9 million. Gold had 67.4% of the chips in play and the total prize money left was $22.2 million, so a proportional chip deal would have guaranteed him approximately 67.4% × $22.2 million ~ $15.0 million, which not only is far more than $9.9 million but is more than he received for winning first place! A proportional chip deal would have been extremely biased in his favor.

Example 4.2.6

On the last hand of Season 1 of *High Stakes Poker*, Barry Greenstein raised to $2500 with A♥ A♣, Sammy Farha reraised to $12,500 with K♣ K♦, Greenstein re-reraised to $62,500, and after pondering the decision for 3 minutes, Farha decided to go all-in for about $180,000. Greenstein quickly called and had Farha covered. Including the blinds and antes, the pot was $361,800. At this point, given only the two players' cards, Greenstein had about an 81.7% chance to win the hand, Farha had about a 17.8% chance to win, and there was approximately a 0.5% chance that the two players would split the pot (which could happen for instance if the board came 34567 or if all board cards were spades). What is the expected number of chips Farha would have after the hand?

Answer   Let X = the number of chips Farha will have after the hand. X will be $0 if Farha loses and $361,800 if Farha wins; X will be $180,900 if they split the pot. Those are the only three possibilities. To calculate $E(X)$, we simply multiply each possible value of X by its probability and add the products:

$$E(X) = (\$0 \times 81.7\%) + (\$361,800 \times 17.8\%)$$
$$+ (\$180,900 \times 0.5\%)$$

$$= \$65,304.90.$$

Incidentally, the board came 6♣ K♥ 8♥ 7♠ 3♦, so Farha won the huge $361,800 pot. Note that in this problem, as in Example 4.2.3, the expected value of X is not close to any value X can actually take.

Example 4.2.7

In a hand from Season 3, Episode 1 of *High Stakes Poker*, with blinds of $300 and $600 plus $100 antes from each of the six players dealt in the hand, Victor Ramdin had J♣ 8♣ and called the $600 big blind before the flop. While two other players folded, William Chen with 10♠ 9♠ and Mike Matusow and Jamie Gold in the small blind and big blind also saw the flop for $600 each. The flop came K♥ J♠ 10♦, Matusow and Gold checked, Ramdin bet $2500, Chen called, and Matusow and Gold folded. The turn brought the 8♦, Ramdin checked, Chen bet $5000, and Ramdin raised all-in for $9500 more. At this point, Chen faced a tough decision and ultimately decided to call. Let X = the amount Chen wins from this pot, that is, X is the entire amount in the pot if Chen wins and X = 0 if Chen loses. Given the cards on the flop and turn and the hole cards of Ramdin and Chen, calculate $E(X)$. Is calling or folding more favorable to Chen in terms of expected value? By how much?

Answer If Chen wins, then X = $600 in antes + $2400 + $5000 + $10,000 + $19,000 = $37,000. Eight cards are assumed known, and of the 44 equally likely remaining possibilities for the river, Chen wins with a Q, 10, or 7. Ten such cards remain, so the probability that one (Q, 10, or 7) will occur is 10/44. Thus $E(X)$ = ($37,000 × 9/44) + ($0 × 35/44) or approximately $8409. Since the bet is $9500, the expected amount Chen has after the hand is maximized if he folds rather than calls, and the corresponding difference is about $9500 − $8409 = $1091.

In the actual hand, the river was the 4♥ and Ramdin won the $37,000 pot. This example illustrates how, if you know your opponents' cards, you can use the concept of expected value to dictate which poker decision to make. For instance, had Chen known for sure what Ramdin had, Chen would have maximized his expected profit by folding, but only by a difference of $1091, which is a small amount in relation to the sizes of the pots in this game. It was obviously a close call and a difficult decision for Chen, but one that we as observers can solve by simple arithmetic if we know both players' cards. The situation is considerably simplified in cases like Example 4.2.7, where the opponent has bet all-in and the decision is whether to call or fold. Formulas for making these types of decisions and some more complex variations are discussed in Section 4.3.

Before we proceed, however, we should note that although there is a good case for maximizing one's *expected* number of chips on every decision, there may be occasions when this myopic strategy is not optimal. For instance, in a cash game, one may desire to keep opponents guessing and convey an image of a maniacal or extremely tight player if this is not the case. Such longer-term image considerations may correctly affect one's decisions beyond the expected number of chips one will have after the current hand.

Similarly, in a tournament that is not winner-take-all, players are paid on how long they last, not based on their numbers of chips, so a player may wish to be somewhat risk averse and avoid confrontations that are slightly favorable in terms of expected value. Indeed, many poker strategy books advise playing very aggressively just before the prize money *bubble*, because many players will fold nearly any hand rather than risk elimination when the payout is so close. Some of these books express a kind of derision toward these players who play very tightly near the bubble, but while playing very aggressively with a medium or large chip stack near the bubble is certainly excellent strategy, extremely conservative play with a small chip stack near the bubble may in fact be optimal as well.

At other stages of non-winner-take-all tournaments, players may similarly seek conservative or aggressive strategies based on other concerns aside from maximizing their expected numbers of chips, especially when the payouts decline very gradually. Some studies (e.g., Kim 2005) suggest that for the steep payout structures currently used in most major tournaments, the myopic rule still leads to optimal or nearly optimal strategy just as in cases of winner-take-all tournaments.

4.3 Pot Odds

Many factors typically come into play when making decisions in Texas Hold'em. Usually one must guess what likely hands the opponents may have, as well as one's own image, whether a bluff might work, and so on. In some cases, however, the decision is based simply on an arithmetic calculation.

Suppose you face only one opponent and this opponent bets (or raises) you, going all-in. Now your only decision is whether to call or fold. What should your chances of winning be for you to correctly call if your goal is to maximize your expected number of chips?

Let b = the *bet* you are asked to call, i.e., the additional amount you would need to put in if you wanted to call. For instance, if you bet 100 and your opponent with 800 left went all-in, then b = 700, assuming you had 700 left. If you only had 400 left after you bet 100, then b = 400.

Let c = the *current* size of the pot you will receive if you call and win the hand at the time when you are making your decision. Note that this includes your opponent's bet. Let p = your *probability* of winning the hand if you call. For simplicity, ignore the chance of a split pot, so that your chance of losing if you call is $1 - p$. Let n = the *number* of chips you currently have and a = the number of chips you will have *after* the hand. If you fold, then you are certain to have n chips after the hand with 100% probability, so in this case $E(a) = n$. If you call, you will either have $n - b$ chips if you lose or $n + c$ chips if you win.

So in this case,

$$E(a) = [(n - b) \times (1 - p) + (n + c) \times p]$$

$$= n(1 - p + p) - b(1 - p) + cp$$

$$= n - b + bp + cp.$$

Thus, if you want $E(a)$ maximized, you should call if and only if $n - b + bp + cp > n$, i.e., if

$$p > b/(b + c). \tag{4.3.1}$$

This ratio, $r = b \div (b + c)$, is the bet to you divided by the size the pot *will be* after you call, and it arises frequently in poker calculations. Note that r does not depend directly on n. However, if your opponent bets an amount greater than n, then $b = n$ rather than the amount the opponent put in, since in such situations you are only wagering (and only stand to win) the number of chips you have. The product cp in Condition 4.3.1 represents what poker players call *equity* in the pot, which is the portion of the pot you *expect* to win, in an expected value sense, assuming no further betting in the hand.

Note that Condition 4.3.1 is only directly relevant under the following very special restrictions:

1. You want to maximize the expected number of chips you will have after the hand.
2. You know or are able to estimate the probability p that you will win the hand.
3. You are only up against one opponent.
4. Your opponent is all-in.
5. The probability of your splitting the pot is zero (or negligible).

Restriction (1) applies quite reasonably to most situations in cash games and winner-take-all tournaments, but may not be reasonable for some other tournaments. Restriction (2) is tricky because estimation of such probabilities typically depends on one's ability to decipher what hand or range of hands one's opponents are likely to have—an ability that in large part differentiates expert poker players from others. In some cases, condition 4.3.1 may readily be modified even when (3), (4), or (5) is violated. Some of these situations are discussed later in this chapter. In most such situations, however, it is easier simply to calculate and compare the expected values directly for each choice as in Example 4.2.5, rather than refer to condition 4.3.1.

Some version of relation 4.3.1 appears in virtually every poker strategy book. Poker players often discuss such calculations not in terms of probabilities, but in terms of odds (discussed in Section 3.3). In the above scenario, in order to maximize your expected number of chips, you should call if the odds of your winning the hand are greater than b/c. Poker players typically express this in a slightly more confusing way, namely that you should call if the odds against you winning the hand are less than the ratio c/b, which they often write as $c{:}b$ and refer to as *pot odds*.

Example 4.3.1

Refer back to the hand between Chen and Ramdin discussed in Example 4.2.7. Compare p with $b/(b + c)$ to verify whether Chen maximizes his expected number of chips by calling or folding.

Answer Here p = 10/44 ~ 22.7%, b = \$9500, and c = \$27,500, so the ratio $r = b/(b + c)$ ~ 25.7%. Since $p < 25.7\%$, we confirm as in Example 4.2.7 that Chen's expected number of chips is maximized by folding rather than calling.

When your opponent is *not* all-in, restriction (4) is violated, and in such cases it can be less clear whether to fold, call, or raise. It may be that if you call the current bet and take the lead later in the hand, then you will win more than the sum $b + c$ in relation 4.3.1 because you may get your opponent to lose more chips later in the hand. Similarly, after you call the current bet, you may face another bet from your opponent during the next betting round, so you may stand to lose more than b. Poker players use the terms *implied odds* and *reverse implied odds*, respectively, in discussing these two possibilities, and the term *pot odds* (or *express odds*) refers to the simple ratio $c{:}b$. For instance, if on the turn you are facing a bet of b, the current size of the pot is c, and you have probability p of winning the hand on the river, in which case you will win an additional d chips from your opponent on the river, then your express odds are $c{:}b$ and your implied odds are $(c + d){:}b$. In terms of probabilities rather than odds, in order to maximize your expected number of chips, you should call if $p \geq b/(b + c + d)$. If all of the above is true and furthermore, if you lose on the river, you will be tricked into losing an additional e chips on the river, then the relevant ratio is instead $(b + e)/(b + c + d)$. Often, however, whether you win or lose additional chips depends on the river card, in which case the simple ratio above does not apply, as in the next example.

Example 4.3.2

In a hand during Season 2, Episode 4 of *High Stakes Poker*, Minh Ly reraised to $11,000 before the flop with K♥ K♦ and Daniel Negreanu called with A♠ J♠. Because of betting from other players, the pot was now $37,300. The flop came 8♠ 7♥ 2♠, Ly bet $15,000, and Negreanu called. The turn was 4♣ and Ly bet $50,000. Given these eight cards, in terms of maximizing expected profits, would it be better for Negreanu to call or fold? Make the following simplifying assumptions.

- If the river were an ace, then Negreanu would bet $50,000 and Ly would fold.
- If the river were a spade, then Negreanu would bet $50,000 and Ly would call.
- If the river were a jack, then Ly would bet $100,000 and Negreanu would call.
- If the river were anything else, then Ly would bet $100,000 and Negreanu would fold.

Answer Assuming he calls, let X represent Negreanu's profit in the hand relative to folding. We will calculate $E(X)$. The current size c of the pot before Negreanu calls is $37,300 + $30,000 + $50,000 = $117,300. Forty-four remaining cards are equally likely to appear on the river. If the river is an ace, then Negreanu simply wins the pot and $X = \$117,300$. If the river is a spade, then Negreanu wins the \$117,300 currently in the pot plus an additional \$50,000 from Ly, so his profit, X, relative to folding, is \$167,300. If the river is a jack, then Negreanu loses $b = \$50,000$ plus an additional \$100,000, so $X = -\$150,000$. If the river is any of the 29 other cards, then $X = -\$50,000$. Thus $E(X) = (3/44 \times \$117,300) + (9/44 \times \$167,300) + (3/44 \times -\$150,000) + (29/44 \times -\$50,000) \sim \$7998 + \$34,220 - \$10,227 - \$32,955 = -\$964$, so Negreanu's expected profit is higher by folding than by calling under these assumptions.

In the actual hand, Negreanu folded, but the players looked at what would have been the river card, and it was J♣.

The hand in Example 4.3.2 illustrates another important complication involved in these calculations of expected profit in no-limit Texas Hold'em. When considering Negreanu's call on the flop, it is tempting to calculate p as Negreanu's probability of winning the hand in a showdown. If, however, Ly is likely to bet \$50,000 on the turn, in which case Negreanu will fold, then p should be calculated as the probability of Negreanu taking the lead on the turn only. Of course, an additional complication is that when playing an actual poker hand, one typically knows neither the opponents' cards nor what their strategies will be on subsequent betting rounds.

When the other restrictions hold but restriction (3) is violated, i.e., you have two or more opponents, the situation is very similar to the direct application of condition 4.3.1, provided you have the smallest number of chips. The relevant calculation is still to compare your probability of winning the hand with the ratio r, where again r is the bet to you divided by the size of the pot you will receive if you call and win.

Example 4.3.3

Consider again the WSOP hand from the beginning of Chapter 1. To recap, Wasicka started the hand with $18 million in chips, Binger with $11 million, and Gold with $60 million. Wasicka had 8♠ 7♠, Binger had A♥ 10♥, and Gold had 4♠ 3♣. Binger raised to 1.5 million before the flop, Wasicka and Gold called, and the flop was 10♣ 6♠ 5♠. Binger bet 3.5 million and Gold raised all-in. Wasicka folded, but if Wasicka had called, Binger would have faced an incredibly interesting decision. If he folds, then all he needs is for Gold to beat Wasicka in the hand, and Binger assures himself of at least second place and an additional $2 million in prize money. However, if he calls and wins the hand, then he has $33 million in chips and an excellent chance for first place. In terms of his expected number of chips, it's an easy decision, especially if we assume he knew what his opponents had. Under this assumption, determine whether Binger maximizes his expected number of chips via calling or folding (1) by directly comparing Binger's expected number of chips if he were to call with his expected number of chips if he folds and (2) by comparing p and r.

Answer

1. If Binger folds, then he is certain to end the hand with his remaining 6 million chips $(11 - 1.5 - 3.5 = 6)$. So, if he folds, then his expected number of chips is simply $100\% \times 6$ million = 6 million. If he calls, then he will either have $33 million in chips if he wins the hand, or 0 if he loses the hand. There are $C(43,2)$ equally likely combinations of turn and river cards, and Binger wins the hand if the turn and river are (a,b), where a and b are any of the 22 non-spade and non-2, 3, 4, 7,

8, 9 cards left in the deck. He also wins if the turn and river are

(a, c) where c = 3♦ or 3♥,

or (a, d) where d = an 8,

or (c, d),

or $(10♠, e)$, where e = 5, 6, A, or 10♦,

or $(A♠, f)$, where f = 10♦, A♣, or A♦,

so p = Binger's probability of winning

$$= [C(22,2) + 22 \times 2 + 22 \times 3 + 2 \times 3 + 1 \times 10 + 1 \times 3] \div C(43,2)$$

$$= 360/903 \sim 39.9\%.$$

Thus, the expected number of chips Binger will have if he calls is (33 million × 39.9%) + (0 × 60.1%) or ~13.2 million chips.

If Binger used a myopic strategy of trying to maximize his expected number of chips, he would obviously call since 13.2 million is far more than 6 million.

2. Here r = 6 million ÷ 33 million or ~18.2%, and p = 39.9% > 18.2%, so calling maximizes his expected number of chips.

Note that although the myopic strategy clearly suggests calling, in reality it would have been a difficult decision, since this is not a winner-take-all tournament where the myopic rule should necessarily be used. If Binger folds, he substantially increases his chance for second place; if he calls, he not only increases his chance for first place but also increases his chance for third place.

If one of your opponents has fewer chips than you have, the situation is more complex, because even if you lose to this opponent, you may win a side pot against your other opponents. Similarly, if restriction (e) is violated, then in addition to the chance of winning the pot, you must take into account the possibility of splitting the pot. In these cases, there is more than one probability to consider, and in such situations it is simpler to avoid relation 4.3.1 and instead compare expected values associated with each choice directly, as in the following example.

Example 4.3.4

Consider the hand from the previous example, but from Paul Wasicka's perspective. To simplify things, suppose Wasicka knew what cards his opponents had and also knew that, if he called after Gold raised all-in, Binger would have called also. If Wasicka wanted to maximize his expected number of chips after the hand, should he have called or folded after Gold went all-in?

Answer If Wasicka calls, four outcomes are possible. First, if Wasicka wins the hand, he wins all 11 million in chips from Binger plus 18 million from Gold, so he ends up with a total of 47 million in chips. Second, if Gold's hand beats Wasicka's, then Wasicka loses all of his chips to Gold. Third, if Binger wins the hand and Wasicka's hand beats Gold's, then Wasicka loses 11 million in chips to Binger but wins 7 million in chips from Gold, so Wasicka ends up with 14 million in chips. Finally, the fourth possibility is that the turn and river combination is (5,6). In this case, Binger wins the hand overall but Gold and Wasicka split the side pot, since the best five-card hands for both Gold and Wasicka consist of the five board cards. In this case Wasicka loses 11 million in chips to Binger and wins 0 from Gold, so Wasicka ends with 7 million in chips.

The probabilities associated with each of these four outcomes must be calculated. The probability of Wasicka winning the hand was calculated in Example 2.4.6 as 486/903 or ~53.82%. Gold's hand beats Wasicka's if the turn and river combination is $(2n, 2n)$, $(2n, a)$, $(3\heartsuit, 3\spadesuit)$, $(3n, b)$, $(7, 7)$, $(7, c)$, or $(2\spadesuit, 3\spadesuit)$. The variable n indicates non-spade; a is not a spade, 2, 4, or 9 (there are $43 - 8 - 3 - 3 - 3 = 26$ such cards remaining); b is not a spade, 2, 3, 4, 8, or 9 (21 of these cards remain); and c is not a spade, 2, 3, 4, 7, or 9 (21 such cards remain). Thus the probability of Gold's hand beating Wasicka's is $[C(3,2) + 3 \times 26 + 1 + 2 \times 21 + C(3,2) + 3 \times 21 + 1] \div C(43,2) = 191/903$ or ~21.15%.

Wasicka's hand loses to Binger's but beats Gold's if the turn and river combination is $(8n, d)$, $(10\spadesuit, e)$, $(A\spadesuit, f)$, $(5n, 5n)$, $(6n, 6n)$, $(5n, g)$, $(6n, g)$, or (g, h), where n means non-spade; d is not a spade, 2, 4, 7, 8, or 9 (20 of these cards remain); e is 5, 6, A, or $10\diamond$; f is $10\diamond$, $A\clubsuit$, or $A\diamond$; and g and h can be any non-spade 10, J, Q, K, or A (there are 12 cards left in this category). The corresponding probability is $[3 \times 20 + 1 \times 10 + 1 \times 3 + C(3,2) + C(3,2) + 3 \times 12 + 3 \times 12 + C(12,2)] \div C(43,2) = 217/903$ or ~24.03%.

Finally, the probability that Wasicka loses to Binger but splits the side pot with Gold is the probability of the turn and river combination coming $(5, 6)$, which is simply $3 \times 3 \div C(43,2) = 9/903$ or ~1.00%.

Observe that the probabilities sum to 1: $53.82\% + 21.15\% + 24.03\% + 1.00\% = 100\%$. Now the expected value of the number of chips Wasicka will have after the hand if he calls can be calculated as $(53.82\% \times 47$ million$) + (21.15\% \times 0) + (24.03\% \times 14$ million$) + (1.00\% \times 7$ million$)$ or ~28.7 million chips. If Wasicka folds, he will, with 100% probability, have 16.5 million chips remaining, so if his goal were to maximize his expected number of chips after the hand, then he clearly should have called.

Wasicka did in fact fold in this situation, after some deliberation and understandable consternation. Obviously, since the tournament was not winner-take-all, the myopic rule would not apply, and in addition, Wasicka may have been thinking that he could outplay his opponents in the future; in such cases, the tournament is not symmetric, the myopic strategy is not necessarily optimal, and one may choose to avoid situations that are slightly favorable in terms of expected value and instead wait for *more* favorable situations in the future. Note, however, that Wasicka did not appear to outplay Gold at the end of the tournament. Ironically, it is not uncommon for players who consider themselves vastly superior in skill to their opponents actually to play worse. Poker skill is difficult to quantify, however, and some ideas on this are discussed in Section 4.4.

The previous examples involve the decision of whether or not to call a bet. Expected value can also be applied to decisions involving whether or not to bluff, as in the following three examples.

Example 4.3.5

Suppose you cannot possibly win in a showdown and are only considering two options: either betting the size of the pot as a bluff or simply checking and giving up. How often does the pot-sized bluff have to work in order to be profitable in the long term?

Answer Suppose c is the current size of the pot, p is the probability that our opponent will fold to a bluff, and x is the number of chips you currently have. If you bluff, your opponent might fold, in which case you will have $x + c$ chips, or your opponent might call, in which case you will have $x - c$ chips. So, the expected number of chips you will have after the hand is $(x + c)p + (x - c)(1 - p) = x + 2cp - c$. If you check, your number of chips is simply x. Therefore you should bet if $2cp - c > 0$, i.e., if $p > \frac{1}{2}$.

Note that the answer in Example 4.3.5 does not depend on c or x. The answer above assumes that the opponent will either call or fold, but sometimes the opponent can raise, leading to another decision about whether to bluff or not. The following example illustrates this.

Example 4.3.6

In the 2014 WSOP Main Event, with just 19 players remaining, Tom Sarra, Jr., had 8♥ 7♥ against Dan Sindelar with K♥ Q♣ on a flop of 6♣ 5♣ J♠. After Sindelar bet and Sarra called, the pot was 1.75 million. The turn was 6♦, both players checked, and when the Q♠ came on the river, Sarra bluffed 560,000, Sindelar raised to 1.475 million, and Sarra boldly reraised to 5.06 million, getting Sindelar to fold. Before Sarra's reraise, the pot was 3.785 million. Sarra then put in 5.004 million more chips as a bluff. How often would a bluff reraise of this size have to work in order to be profitable in the long term?

Answer Let M stand for *million*. Suppose Sarra's bluff reraise works and gets his opponent to fold with probability p, in which case Sarra gains $3.785M$ chips as a result, or his opponent calls with probability $1 - p$, in which case Sarra will lose $5.004M$ chips because of the bluff reraise. The reraise has positive expected value for Sarra if

$$p(3.785M) > (1 - p)(5.004M),$$
i.e., if $p > 5.004M/8.789M \sim 56.93\%$.

Example 4.3.7

On day 4 of the 2015 WSOP Main Event, Phil Hellmuth, Jr., made a great bluff. Jae Kim had 8♣ 8♦ on the button and called 13,000. Hellmuth had 7♦6♣ in the small blind, raised to 40,000, and Kim called. The pot was 108,000 with antes and blinds. The flop was J♥ 3♥ 9♦ and both players checked. The turn was 5♠, and Hellmuth bet 38,000. Kim called and the pot was 184,000. The river was the 3♠. Hellmuth bluffed 70,000, leaving himself with just 23,000 chips left over, and Kim folded, allowing Hellmuth to win the 254,000 chip pot. How often would Hellmuth's bluff need to have worked in order for it to give him a positive expected return?

Answer If the bluff works, it wins 184,000 chips for Hellmuth, and if it fails it costs him 70,000 chips. So, if Hellmuth's bluff works with probability p, then in order for it to yield positive expected return for Hellmuth, we need

$$184,000 \, p > 70,000 \, (1 - p),$$
$$\text{i.e., } p > 70,000 / 254,000 \sim 27.56\%.$$

Example 4.3.8

Looking at the previous example from Jae Kim's perspective, Kim was unsure whether to fold or to call Hellmuth's river bet. With at least what probability would calling have had to be correct in order for Kim to maximize his long-term expected number of chips by calling on the river rather than folding?

Answer The bet was b = 70,000, and the pot size at the time of Kim's decision on the river was c = 184,000 + 70,000 = 254,000. By relation 4.3.1, Kim should call if he will win with frequency at least $b/(b + c)$ = 70,000/324,000 ~ 21.60%.

4.4 Luck and Skill in Texas Hold'em

The determination of whether Texas Hold'em is primarily a game of luck or skill has recently become the subject of intense legal debate. The terms *luck* and *skill* are extremely difficult to define. Surprisingly, rigorous definitions of these terms seldom appear in books and journal articles on game theory. A few articles have defined *skill* in terms of the variance in results among different players, with the idea that players should perform more similarly if a game is mostly based on luck, but their results may differ more substantially if a game is based on skill. Another definition of *skill* is the extent to which players can improve. Poker does indeed involve a significant amount of potential for improvement (Dedonno and Detterman, 2008). Others have defined *skill* in terms of the variation in a given player's results, since less variation would indicate that fewer repetitions are necessary to determine the statistical significance of a long-term edge in the game and, hence, the sooner one can establish that average profits or losses are primarily due to skill rather than short-term luck.

These definitions are obviously extremely problematic for various reasons. First, both definitions rely on repeated playing of the game in question before even a crude assessment of luck or skill may be made. More importantly, there are many contests of skill wherein the differences between players are small, or where one's results vary wildly. For instance, in Olympic trials of the 100-meter sprints, the differences between finishers are typically quite small, often just hundredths of a second. This hardly implies that the results are based on luck. In other sporting events, for example pitching in baseball, an individual's results may vary widely from one day to another, but that hardly means that luck plays a major role.

To quantify the amount of luck or skill in a particular game of poker, one possibility may be to define *luck* as equity gained when cards are dealt by the dealer, and *skill* as expected profit gained by a player's actions during betting rounds. (Recall that *equity* was defined in Section 4.3 as the product cp.) You might gain expected profit during a hand by several actions:

- ■ The cards dealt by the dealer (whether the players' hole cards or the flop, turn, or river) give you a greater chance of winning a hand in a showdown, thus increasing your equity in the pot.

- The size of the pot is increased while your chance to win the hand in a showdown is better than those of your opponents.
- By betting, you get others to fold and thus increase your probability of winning the pot.

Certainly, anyone would characterize the first case as luck, unless perhaps one believes in ESP or time travel. Thus, it may be possible to estimate skill in poker by looking at the second and third cases above. That is, we may view skill as the expected profit gained during the betting rounds, whereas luck is expected profit gained simply by dealing the cards. Both are easily quantifiable, and one may dissect a particular poker game and analyze how much expected profit each player gained due to luck or skill.

There are obvious objections to these definitions. First, why calculate expected profit or equity in a pot assuming no future betting? The assumption of no future betting may seem absurdly simplistic and unrealistic. On the other hand, unlike *implied equity*, which accounts for betting on future betting rounds, ordinary equity is unambiguously defined and easy to compute. A second objection is that situations can occur where a terrible player may gain expected profit during betting rounds against even the greatest player in the world and attributing such gains to *skill* may be objectionable. For instance, in heads-up Texas Hold'em, if the two players are dealt AA and KK, one would expect the player with KK to put a great number of chips in while way behind. This situation seems more like bad luck for the player with KK than a deficit in skill. One possible response to this objection is that skill is difficult to define, and in fact most poker players, probably due to their huge and fragile egos, tend to attribute nearly all losses to bad luck. Almost anything can be attributed to luck if the definition of luck is general enough. Even if a player makes an amazingly skillful poker play, such as folding a very strong hand because of an observed tell or betting pattern, one could argue that the player was lucky to observe the tell or even that he was lucky to have been born with the ability to discern the tell. On the other hand, situations like the AA versus KK example truly do seem like bad luck. It is difficult to think of any remedy to this problem. It may be that *skill* is too strong a word, and that when analyzing hands in terms of equity, one should perhaps instead refer to expected profit gained *during betting rounds* rather than expected profit gained *due to skill*. We will nevertheless use the word *skill* in what follows.

Example 4.4.1

On day 4 of the WSOP Main Event in 2015, with blinds of 5000 and 10,000 and antes of 1000 from eight players, after Ryan D'Angelo raised to 22,000 with A♦ K♠, Daniel Negreanu called 17,000 more from the small blind with A♠ 7♠, and Fernando Perez called from the small blind with 3♥ 2♥. The pot was 74,000. The flop came 3♠ 10♣ 9♠ and everyone checked. The turn was 2♠, giving Negreanu the nuts and giving Perez two pairs. Negreanu bet 35,000, Perez raised to 105,000, D'Angelo folded, Negreanu reraised to 250,000, and Perez called. The pot was now 574,000. The river was 5♣, Negreanu bet all-in for 359,000, and Perez folded. How much expected profit did Negreanu gain (1) due to luck on the turn, (2) due to skill on the turn, (3) due to luck on the river, and (4) due to skill on the river?

Answer

1. Before the turn was revealed,
   $P(\text{Negreanu wins})$

   $= P(\text{spade on turn or river}) + P(77)$

   $+ P(7x) + P(J^*8^*) + P(8^*6^*),$

where x is a non-spade card that is not a 2, 3, or K and * denotes non-spades,

$$= [8/45 + 8/45 - C(8,2)/C(45,2)]$$
$$+ C(3,2)/C(45,2) + 3 \times 34/C(45,2)$$
$$+ 3 \times 3/C(45,2) + 3 \times 3/C(45,2)$$

~ 45.15%.

(There was also a small [1/165] chance of a split pot with 10*9* but we will ignore that here.) When the turn was revealed, the probability that Negreanu would win, assuming nobody folded, was the probability that a 3 or 2 would not come on the river, which is 41/44 ~ 93.18%. Thus Negreanu's equity increased from 45.15% × 74,000 to 93.18% × 74,000 due to luck on the turn, an increase of 35,542.2 chips.

2. During the betting on the turn, the pot increased from 74,000 chips to 574,000 chips. Thus Negreanu's equity increased from 93.18% × 74,000 = 68,953.2 to 93.18% × 574,000 = 534,853.2, for an increase in equity of 465,900. The cost to Negreanu on the turn was 250,000 chips, so his increase in expected *profit* on the turn due to skill was 465,900 − 250,000 = 215,900 chips.

3. On the river, Negreanu went from having a 93.18% chance of winning the hand in a showdown to 100%, so his equity increased from 534,853.2 to 574,000, for an increase of 39,146.8 due to luck.

4. The river betting did not increase Negreanu's profit, so Negreanu gained 0 due to skill on the river.

Example 4.4.2

Consider the 2015 WSOP Main Event hand between Cloud and Hellmuth in Example 3.1.5, where Cloud raised to 15,000 with A♣ A♠,

Hellmuth called with A♥ K♠, Daniel Negreanu called from the big blind with 6♦ 4♥, and the flop came K♣ 8♥ K♥. Before the flop, the pot was 57,000 chips, and the probabilities shown on ESPN's broadcast of winning the hand in a showdown at this point were 74% for Cloud, 19% for Negreanu, and only 6% for Hellmuth. (The probabilities only add up to 99% because of an approximately 1% chance of a split pot.) After the flop, all three players checked, the turn was the J♥, Negreanu checked, Cloud bet 15,000, Hellmuth called, and Negreanu folded. The river was 7♠, Cloud checked, Hellmuth bet 37,000, and Cloud called. How much expected profit did Hellmuth gain due to luck and how much due to skill (1) on the flop, (2) on the turn, and (3) on the river?

Answer

1. Before the flop was revealed, Hellmuth's equity was 6% × 57,000 = 3420 chips. After the flop was dealt, the only way Hellmuth could have lost in a showdown would have been if the turn or river contained the A♦ without the K♦, which, given the six cards belonging to the players and the three cards on the flop, had a probability of (1 × 41)/ $C(43,2)$ = 4.54%, so Hellmuth's equity suddenly increased to 95.46% × 57,000 = 54,412.2 chips. Thus on the flop Hellmuth gained 54,412.2 − 3420 = 50,992.2 chips in equity due to luck. There was no betting on the flop so Hellmuth gained 0 expected profit due to skill on the flop.

2. When the turn was dealt, Hellmuth's probability of winning in a showdown increased to 41/42 ~ 97.62%, so his equity increased from 54,412.2 to 97.62% × 57,000 = 55,643.4, for an increase in expected profits of 1231.2 due to luck on the turn.

During the betting on the turn, Hellmuth and Cloud each put 15,000 chips in the pot, so Hellmuth's expected return increased by 97.62% × 30,000 = 29,286 chips, but he put 15,000 chips into the pot on the turn, so his expected *profit* on the turn due to skill was 29,286 − 15,000 = 14,286 chips.

3. After the turn betting, the pot was 87,000 chips. When the 7♠ was revealed on the river, Hellmuth's equity increased from 97.62% × 87,000 = 84,929.4 to 100% × 87,000, for an increase of 2070.6 chips due to luck. Hellmuth's expected profit gained due to skill on the river is simply 37,000 chips: the pot size increased by 74,000, while Hellmuth had a 100% chance of winning, but the cost to Hellmuth was 37,000, so his profit was 37,000.

Example 4.4.2 shows what one might consider a problem with defining skill and luck in terms of expected profit and equity. Clearly Hellmuth got extremely lucky. The analysis here attributes 50,992.2 + 1231.2 + 2070.6 = 54,294 of his profits to luck. However, it also credits Hellmuth with 0 + 14,286 + 37,000 = 51,286 chips in profit due to skill. One issue with the definitions of *luck* and *skill* proposed here is that luck and skill will tend to be *correlated*: players who are lucky enough to get better cards than their opponents will typically bet when they are ahead and thus gain in skill as well. We will discuss correlation further in Section 7.1.

The extended example below is intended to illustrate the division of luck and skill in a game of Texas Hold'em. It took place at the end of a tournament on *Poker After Dark* televised in October 2009. Dario Minieri and Howard Lederer were the final two players. Since this portion of the tournament involved only these two players, and since most of the hands were televised, this example lets us attempt to parse out how much of Lederer's win was due to skill and how much to luck.

Technical note: Before we begin, we must clarify a few potential ambiguities. There is some ambiguity in the definition of expected profit before the flop, since the small and big blind put in different numbers of chips. The definition used here is the equity a player would have in the pot after calling minus cost, assuming the big blind and small blind call as well, or the (negative) profit a player would have by folding, whichever is greater. For example, in heads-up Texas Hold'em with blinds of 800 and 1600, the pre-flop expected profit for the big blind is $3200p - 1600$, and $max\{3200p - 1600, -800\}$ for the small blind, where p is the probability of the big blind winning the pot in a showdown. Define increases in the size of the pot as relative to the big blind, i.e., increasing the pot size by calling preflop does not count as skill. The probability p of winning the hand in a showdown was obtained using the odds calculator at cardplayer.com, and the probability of a tie is divided equally between the two players in determining p.

Example 4.4.3

Below are summaries of all 27 hands shown on *Poker After Dark* in October 2009 for Dario Minieri and Howard Lederer in the heads-up segment of the tournament, with each hand's gains and losses in expected profit categorized as luck or skill. Each hand is analyzed from Minieri's perspective, i.e., *skill* −100 means 100 chips of expected profit gained by Lederer during a betting round. The question we seek to address is how much of Lederer's win was due to skill and how much of it was due to luck?

Answer   For example, here is a detailed breakdown of hand 4 in which the blinds were 800 and 1600, Minieri was dealt A♣ J♣, Lederer had

A♥ 9♥, Minieri raised to 4300 and Lederer called. The flop was 6♣ 10♠ 10♣, Lederer checked, Minieri bet 6500, and Lederer folded.

1. Pre-flop dealing (luck): Minieri +642.08. Minieri was dealt a 70.065% probability of winning the pot in a showdown so his increase in expected profit is

    70.065% × 3200 − 1600 = 642.08 in chips.

    Lederer was dealt a 29.935% probability to win the pot in a showdown, so his increase in expected profit is

    29.935% × 3200 − 1600 = −642.08.

2. Pre-flop betting (skill): Minieri +1083.51. The pot was increased to 8600. Since 8600 − 3200 = 5400, Minieri had 70.065% × 5400 = 3783.51 additional equity but paid an additional 2700, so his expected profit due to betting was 3783.51 − 2700 = 1083.51. Correspondingly, Lederer's expected profit due to betting was −1083.51 since 29.935% × 5400 − 2700 = −1083.51.

3. Flop dealing (luck): Minieri +1362.67. After the flop was dealt, Minieri's probability of winning the 8600-chip pot in a showdown increased from 70.065% to 85.91%. Because of luck, he increased his equity by (85.91% − 70.065%) × 8600 = 1362.67 chips.

4. Flop betting (skill): Minieri +1211.74. Because of betting on the flop, Minieri's equity went from 85.91% of the 8600 chip pot to 100% of the pot so he increased his equity by (100% − 85.91%) × 8600 = 1211.74 chips.

So during the hand, due to luck, Minieri increased his equity by 642.08 + 1362.67 = 2004.75 chips. Due to skill, he increased his expected profit by 1083.51 + 1211.74 = 2295.25 chips. Note that the total = 2004.75 + 2295.25 = 4300, the number of chips Minieri actually won from Lederer in the hand.

Note that before the heads-up battle began, the broadcast reported that Minieri had 72,000 chips and Lederer had 48,000. Minieri must have won some chips in hands that were not televised because the grand total showed Minieri losing about 74,500 chips.

(Blinds 800/1600)

Hand 1. Lederer A♣ 7♠, Minieri 6♠ 6♦. Lederer 43.535%, Minieri 56.465%. Lederer raises to 4300. Minieri raises to 47,800. Lederer folds. Luck +206.88. Skill +4093.12.

Hand 2. Minieri 4♠ 2♦, Lederer K♠ 7♥. Minieri 34.36%, Lederer 65.64%. Minieri raises to 4300; Lederer raises all-in for 43,500; Minieri folds. Luck −500.48. Skill −3799.52.

Hand 3. Lederer 6♥ 3♦, Minieri A♦ 9♣. Lederer 34.965%, Minieri 65.035%. Lederer folds in the small blind. Luck +481.12. Skill +318.88.

Hand 4. Minieri A♣ J♣, Lederer A♥ 9♥. Minieri 70.065%, Lederer 29.935%. Minieri raises to 4300; Lederer calls 2700. Flop 6♣ 10♠ 10♣. Minieri 85.91%, Lederer 14.09%. Lederer checks; Minieri bets 6500; Lederer folds. Luck +2004.75. Skill +2,295.25.

Hand 5. Lederer 5♠ 3♥, Minieri 7♦ 6♠. Lederer 35.765%, Minieri 64.235%. Lederer folds in the small blind. Luck +455.52. Skill +344.48.

Hand 6. Minieri K♥ 10♣, Lederer 5♦ 2♦. Minieri 61.41%, Lederer 38.59%. Minieri raises to 3200; Lederer raises to 9700; Minieri folds. Luck +365.12. Skill −3565.12.

Hand 7. Minieri 10♦ 7♠, Lederer Q♣ 2♥. Minieri 43.57%, Lederer 56.43%. Minieri raises to 3200; Lederer calls 1600. Flop 8♠ 2♠ Q♥. Minieri 7.27%, Lederer 92.73%. Lederer checks; Minieri bets 3200; Lederer calls. Turn 4♦.

Minieri 0%, Lederer 100%. Lederer checks; Minieri bets 10,000; Lederer calls. River A♥. Lederer checks, Minieri checks. Luck −205.76 − 2323.20 − 930.56 = −3459.52. Skill −205.76 − 2734.72 − 10,000 = −12,940.48.

Hand 8. Lederer 7♣ 2♦, Minieri 9♣ 4♦. Minieri 64.28%, Lederer 35.72%. Lederer folds. Luck +456.96. Skill +343.04.

Hand 9. Minieri 4♠ 2♣, Lederer 8♥ 7♦. Minieri 34.345%, Lederer 65.655%. Minieri raises to 3200; Lederer calls 1600. Flop 3♦ 9♥ J♥. Minieri 22.025%, Lederer 77.975%. Lederer checks; Minieri bets 4800; Lederer folds. Luck −500.96 − 788.48 = −1289.44. Skill −500.96 + 4990.40 = +4489.44.

Hand 10. Lederer K♠ 5♠, Minieri K♥ 7♣. Minieri 59.15%, Lederer 40.85%. Lederer calls 800; Minieri raises to 6400; Lederer folds. Luck +292.80. Skill +1307.20.

Hand 11. Minieri A♥ 8♥, Lederer 6♥ 3♠. Minieri 66.85%, Lederer 33.15%. Minieri raises to 3200. Lederer folds. Luck +539.20. Skill +1060.80.

Hand 12. Lederer A♦ 4♦. Minieri 7♦ 3♥. Minieri 34.655%, Lederer 65.345%. Lederer raises to 4300; Minieri raises to 11,500; Lederer folds. Luck −491.04. Skill +4791.04.

Hand 13. Minieri 6♣ 3♣, Lederer K♠ 6♠. Minieri 29.825%, Lederer 70.175%. Minieri raises to 4800; Lederer calls 3200. Flop 5♥ J♣ 5♣. Minieri 47.425%, Lederer 52.575%. Lederer checks, Minieri bets 6000; Lederer folds. Luck −645.60 + 1689.60 = +1044. Skill −1291.20 + 5047.20 = +3756.

Hand 14. Lederer 7♦ 5♠, Minieri 8♦ 5♦. Minieri 69.44%, Lederer 30.56%. Lederer calls 800, Minieri checks. Flop K♥ 10♠ 8♣. Minieri 94.395%, Lederer 5.605%. Minieri checks; Lederer bets 1800; Minieri calls. Turn 7♠. Minieri 95.45%, Lederer 4.55%. Minieri checks, Lederer checks. River 6♥. Check, check. Luck +622.08 + 798.56 + 71.74 + 309.40 = +1801.78. Skill 0 + 1598.22 + 0 + 0 = +1598.22.

(Blinds 1000/2000)

Hand 15. Minieri 9♦ 5♠, Lederer A♥ 5♦. Minieri 26.755%, Lederer 73.245%. Minieri calls 1000; Lederer raises to 7000; Minieri raises to 14,000; Lederer calls 7000. Flop 10♠ Q♦ 6♥. Minieri 15.35%, Lederer 84.65%. Lederer checks; Minieri bets 14,000; Lederer folds. Luck −929.80 − 3193.40 = −4123.20. Skill −5578.80 + 23,702 = +18,123.20.

Hand 16. Lederer 5♠ 5♥, Minieri A♣ J♦. Minieri 46.085%, Lederer 53.915%. Lederer calls 1000; Minieri raises to 26,800; Lederer calls all-in. The board is 3♠ 9♠ K♠ 10♦ 9♦. Luck −156.60 − 24,701.56 = −24,858.16. Skill −1941.84.

Hand 17. Minieri K♣ 10♣, Lederer 7♦ 5♦. Minieri 62.22%, Lederer 37.78%. Minieri raises to 5000; Lederer calls 3000. Flop J♠ J♦ 4♠. Minieri 69.90%, Lederer 30.10%. Check check. Turn 8♠. Minieri 77.27%, Lederer 22.73%. Lederer bets 6000; Minieri folds. Luck +488.80 + 768 + 737 = +1993.80. Skill +733.20 + 0 − 7727 = −6993.80.

Hand 18. Lederer 5♠ 5♣, Minieri 10♠ 6♥. Minieri 46.12%, Lederer 53.88%. Lederer calls 1000; Minieri checks. Flop 7♣ 8♣ Q♥. Minieri 38.235%, Lederer 61.765%. Minieri checks; Lederer bets 2000; Minieri calls. Turn J♥. Minieri 22.73%, Lederer 77.27%. Minieri bets 4000; Lederer folds. Luck −155.20 − 315.40 − 1240.40 = −1711. Skill 0 − 470.60 + 6181.60 = +5711.

Hand 19. Lederer K♥ 5♠, Minieri K♣ 10♦. Minieri 73.175%, Lederer 26.825%. Lederer raises to 5000; Minieri calls 3000. Flop J♦ 8♥ 10♥.

Minieri 92.575%, Lederer 7.425%. Check, check. Turn 5♦. Minieri 95.45%, Lederer 4.55%. Minieri bets 6000; Lederer folds. Luck +927 + 1940 + 287.50 = +3154.50. Skill +1390.50 + 0 + 455 = +1845.50.

Hand 20. Minieri 7♣ 2♠, Lederer Q♠ 9♠. Minieri 30.205%, Lederer 69.795%. Minieri raises to 6000. Lederer calls 4000. Flop A♦ A♠ Q♦. Minieri 1.165%, Lederer 98.835%. Lederer checks; Minieri bets 6000; Lederer calls. Turn J♣. Minieri 0%, Lederer 100%. Lederer checks; Minieri bets 14,000; Lederer raises to 35,800; Minieri folds. Luck −791.80 − 3484.80 − 279.60 = −556.20. Skill −1583.60 − 5860.20 − 14,000 = −21,443.80.

Hand 21. Minieri 10♥ 3♦, Lederer Q♥ J♠. Minieri 30.00%, Lederer 70.00%. Minieri calls 1000; Lederer checks. Flop 8♠ 4♥ J♣. Minieri 4.34%, Lederer 95.66%. Lederer checks; Minieri bets 2000; Lederer raises to 7500; Minieri raises to 18,500; Lederer raises all-in; Minieri folds. Luck −800 − 1026.40 = −1826.40. Skill 0 − 18,673.60 = −18,673.60.

Hand 22. Lederer A♠ 2♦, Minieri 5♣ 3♥. Minieri 42.345%, Lederer 57.655%. Lederer calls 1000. Minieri checks. Flop K♠ 10♣ 3♠. Minieri 80.10%, Lederer 19.90%. Check, check. Turn Q♠. Minieri 65.91%, Lederer 34.09%. Check; Lederer bets 2000; Minieri folds. Luck −306.20 + 1510.20 − 567.60 = +636.40. Skill 0 + 0 − 2636.40 = −2636.40.

(Blinds 1500/3000)

Hand 23. Minieri 7♥ 7♣, Lederer 8♦ 3♦. Minieri 68.175%, Lederer 31.825%. Minieri all-in for 21,700; Lederer folds. Luck +1090.50. Skill +1909.50.

Hand 24. Minieri Q♥ 5♥, Lederer 8♦ 5♦. Minieri 68.37%, Lederer 31.63%. Minieri all-in for 26,200; Lederer folds. Luck +1102.20. Skill +1897.80.

Hand 25. Lederer 9♣ 3♣, Minieri 5♦ 2♦. Minieri 40.63%, Lederer 59.37%. Lederer folds. Luck −562.20. Skill +2060.20.

Hand 26. Minieri 10♣ 2♠, Lederer 7♣ 7♥. Minieri 29.04%, Lederer 70.96%. Minieri folds. Luck −1257.60. Skill −242.40.

Hand 27. Lederer Q♣ 9♣, Minieri A♣ 5♠. Minieri 55.37%, Lederer 44.63%. Lederer all-in for 29,200. Minieri calls. Board 7♣ 6♣ 10♠ Q♠ 6♦. Luck +322.20 − 32,336.08 = −32,013.88. Skill +2813.88.

Grand totals: Luck −61,023.59. Skill −13,478.41.

Overall, although Lederer's gains were primarily (about 81.9%) due to luck, Lederer also gained more expected profit due to skill than Minieri. On the first 19 hands, Minieri actually gained 20,836.41 in expected profit due to skill and appeared to be outplaying Lederer. On hands 20 and 21, however, Minieri tried two huge unsuccessful bluffs, both on hands (especially hand 20) where he should probably have strongly suspected that Lederer would be likely to call. On those two hands combined, Minieri lost 40,117.40 in expected profit due to skill. Although Minieri played very well on every other hand, all those good plays could not overcome the huge loss of expected profit due to skill in hands 20 and 21.

It is important to note that the player who gains the most expected profit due to skill does not always win. In the first 19 hands of this example, for instance, Minieri gained 20,836.41 chips in expected profit attributed to skill, but because of bad luck, he *lost* a total of 2800 chips over these 19 hands. The bad luck Minieri suffered on hand 16 negated most of his gains due to skillful play. A common misconception is that luck will ultimately balance out, i.e., that one's total good luck will eventually exactly equal one's total bad luck, but this is not true. Assuming one plays the same game repeatedly and independently, and assuming the expected value of equity gained due to luck is 0 (which seems reasonable), then the *average* equity gained by luck per hand will ultimately converge to 0. This is the law of large numbers, discussed further in Section 7.4. It does not imply that one's *total* equity gained by luck will converge to 0, however. Potential misconceptions about the law of large numbers and arguments about possible overemphasis on equity and expected value are discussed in Section 7.4.

To conclude this section, a clear illustration of the potential pitfalls of analyzing a hand purely based on equity is a hand from Season 7 of *High Stakes Poker*. With blinds of $400 and $800 plus $100 antes from each of the eight players, after Bill Klein straddled for $1600, Phil Galfond raised to $3500 with Q♠ 10♥, Robert Croak called in the big blind with A♣ J♣, Klein called with 10♠ 6♠, and the other players folded. The flop came J♠ 9♥ 2♠, giving Croak top pair, Klein a flush draw, and Galfond an open-ended straight draw. Croak bet $5500; Klein raised to $17,500; and Galfond and Croak called. At this point, it is tempting to compute Klein's probability of winning the hand by computing the probability of exactly one more spade coming on the turn and river without making a full house for Croak, or the turn and river including two 6s, or a 10 and a 6. This would yield a probability of $[(8 \times 35 - 4 - 4) + C(3,2) + 2 \times 3] \div C(43,2) = 281/903 \sim 31.12\%$. Klein could also split the pot with a straight if the turn and river were KQ or Q8 without a spade, which has a probability of $[3 \times 3 + 3 \times 3] \div C(43,2) = 18/903 \sim 1.99\%$. These seem to be the combinations Klein needs, and one would not expect Klein to win the pot with a random turn and river combination not on this list, and especially not if the turn and river contain a king and a jack with no spades. However, look at what actually happened.

The turn was the K♣, giving Galfond a straight, and Croak checked; Klein bet $28,000; Galfond raised to $67,000; Croak folded; and Klein called. The river was J♥, Klein bluffed $150,000, and Galfond folded, giving Klein the $348,200 pot!

4.5 Variance and Standard Deviation

Expected value is incredibly useful but only provides limited information about the distribution of a random variable. One is typically also interested in how spread out the distribution of the random variable is, and the variance and standard deviation are indicators of this spread. As a simple example, if one were to wager $10 or $1000 on the flip of a coin, the *expected* profit would be 0 either way (assuming the coin is fair), but the two situations are critically different in that the variability in profit will obviously be higher with the $1000 wager.

For a discrete random variable X, the *variance* of X, which is often abbreviated var(X) or $V(X)$ or s^2, is defined as the expected *squared* difference between X and its expected value. That is, $E(X - \mu)^2 = \Sigma(b - \mu)^2 f(b)$, where the sum is over all values, b, that X may take, and where μ and $f(b)$ are the expected value and probability mass function of X, respectively. Note that var(X) = $E(X - \mu)^2$ = $E(X^2 - 2\mu X + \mu^2) = E(X^2) - \mu^2$.

The *standard deviation* (*SD*) of X is simply the square root of var(x). The variance of X is somewhat difficult to interpret directly, whereas the *SD* can generally be interpreted as the amount by which a particular value of X would *typically* differ from μ.

The definitions of *variance* and *SD* may seem unnecessarily complicated, since we could alternatively measure the amount by which a particular value typically deviates from μ simply by looking at $\Sigma|b - \mu| f(b)$, i.e., the long-term average of the sizes of the deviations from μ. This quantity, $\Sigma|b - \mu| f(b)$, is called the *mean absolute deviation* (MAD), and is rarely used despite its ease of interpretation. The *SD* is interpreted similarly to the MAD, as one may prove mathematically that the *SD* is always at least as big as the MAD and is typically only slightly bigger unless the distribution of the random variable is highly skewed. The *SD* is far more frequently used, largely because of historical convention and because it is often easier to derive its properties.

Example 4.5.1

Consider the hand from Example 3.3.2, where Daniel Negreanu and Eli Elezra went all-in on the flop and the pot was a total of $214,800. Negreanu had A♦ 10♥, Elezra had 6♣ 8♣, and the flop was 6♠ 10♠ 8♥. Let X be the amount of the pot Negreanu wins after the hand. Given their cards and the flop, calculate $E(X)$, $var(X)$, and $SD(X)$, the standard deviation of X.

Answer X = 0 if Elezra wins the hand, X = $214,800 if Negreanu wins the hand, and X = $107,400 if they split the pot, which would occur if the turn and river are (7,9), which has a probability of $4 \times 4 \div C(45,2)$ ~ 1.6%. The probability that Negreanu wins is the chance that the turn and river are (A,A), (10,10), (A,10), (A,b), (10,b), (b,b), (10,6), or (10,8), where b is any non-6, 8, 10, A, and (b,b) means any pair of non-6, 8, 10, A cards. The probability that Negreanu wins = $[C(3,2) + C(2,2) + 3 \times 2 + 3 \times 36 + 2 \times 36 + 9 \times C(4,2) + 2 \times 2 + 2 \times 2] \div C(45,2)$ ~ 25.5%. Thus, the probability that Elezra wins is ~100% − 1.6% − 25.5% = 72.9%.

So, $E(X) = (\$0 \times 72.9\%) + (\$214{,}800 \times 25.5\%) + (\$107{,}400 \times 1.6\%) = \$56{,}492.40$. $E(X^2) = (\$0^2 \times 72.9\%) + (\$214{,}800^2 \times 25.5\%) + (\$107{,}400^2 \times 1.6\%) = \$11{,}950{,}011{,}360$. Thus, $var(X) = E(X^2) - [E(X)]^2 = \$11{,}950{,}011{,}360 - 56{,}492.40^2 = \$8{,}758{,}620{,}102$, and $SD(X) = \sqrt{var(X)}$ or ~$93,587.5.

Note that, if a and b are constants, $E(X) = \mu$, and $Y = aX + b$, then

$$\text{var}(Y) = E[(aX + b)^2] - [E(aX + b)]^2$$

$$= E[a^2 X^2 + 2abX + b^2] - [a\mu + b]^2$$

$$= a^2 E(X^2) + 2ab\mu + b^2 - [a^2\mu^2 + 2ab\mu + b^2]$$

$$= a^2[E(X^2 - \mu^2)]$$

$$= a^2\text{var}(X).$$

Taking square roots, we see that the standard deviation of Y is a times the standard deviation of X. Thus, as mentioned in Section 4.2, multiplying the potential wins and losses by a factor of a amounts to multiplying the expected value and standard deviation of one's profits by a. Note that the additive shift b does not affect the standard deviation.

The standard deviation of no-limit Texas Hold'em is deceptively large. From personal experience and communication with other players, the SD of one's hourly profit in a casino is typically in the range of 20 to 40 times the big blind. Thus, playing no-limit Texas Hold'em is truly gambling in the sense that big wins and big losses arise frequently. The SD can be even higher for tournament play, where one will often incur a relatively small loss of the tournament buy-in fee and occasionally score a huge payout. Thus the variabilities in tournament results are typically enormous.

Example 4.5.2

The No-Limit Hold'em Nationally Televised Regional Championship at Harrah's in Atlantic City, December 19–22, 2010, had 136 participants who paid $10,000 each to enter. Only the last

15 players remaining received payouts. The payouts (rounded to the nearest thousand) were as follows:

Place	Payout ($)
1	358,000
2	221,000
3	160,000
4	117,000
5	88,000
6	67,000
7	52,000
8	41,000
9	32,000
10–12	26,000
13–15	22,000

These were the payouts rather than profits. Thus, 15th place would *profit* $22,000 − $10,000 = $12,000. Suppose you pay $10,000 to enter and have essentially the same strategy as all other players, so you are equally likely to end up in 1st, 2nd, 3rd ... or 136th place. Let X denote your *profit* from the tournament. What are $E(X)$ and $SD(X)$?

Answer $E(X) = (\$348,000 \times 1/136) + (\$211,000 \times 1/136) + (\$150,000 \times 1/136) + (\$107,000 \times 1/136) + (\$78,000 \times 1/136) + (\$57,000 \times 1/136) + (\$42,000 \times 1/136) + (\$31,000 \times 1/136) + (\$22,000 \times 1/136) + (\$16,000 \times 3/136) + (\$12,000 \times 3/136) + (-\$10,000 \times 121/136) = -\$588.24$. Thus, the casino takes on average $588.24 per player as a fee to run the tournament.

$E(X^2) = (\$348,000^2 \times 1/136) + (\$211,000^2 \times 1/136) + (\$150,000^2 \times 1/136) + (\$107,000^2 \times 1/136) + (\$78,000^2 \times 1/136) + (\$57,000^2 \times 1/136) + (\$42,000^2 \times 1/136) + (\$31,000^2 \times 1/136) + (\$22,000^2 \times 1/136) + (\$16,000^2 \times 3/136) + (\$12,000^2 \times 3/136) + (-\$10,000)^2 \times 121/136 = \$1,479,529,412$.

$$\text{Thus, var}(X) = E(X^2) - (E(X)^2)$$

$$= \$1,479,529,412 - (-\$588.24)^2 \sim \$1,479,183,386.$$

$$SD(X) = \sqrt{\$1,479,183,386} \sim \$38,460.$$

Note that the standard deviation of profit is several times larger than the tournament entry fee. This large standard deviation is typical of poker tournaments, and we will look at the impact of this on checking one's results in Chapter 7.

4.6 Markov and Chebyshev Inequalities

The *Markov inequality* states that for any *non-negative* random variable X and any constant $c > 0$,

$$P(X \geq c) \leq E(X)/c.$$

Proof. If X is discrete and non-negative,

$$
\begin{aligned}
E(X) &= \sum_{b} bP(X = b) \\
&= \sum_{b<c} bP(X = b) + \sum_{b \geq c} bP(X = b) \\
&\geq \sum_{b \geq c} bP(X = b) \\
&\geq \sum_{b \geq c} cP(X = b) \\
&= c\sum_{b \geq c} P(X = b) \\
&= cP(X \geq c).
\end{aligned}
$$
■

Example 4.6.1

The 2010 WSOP $1000 no-limit Texas Hold'em event had 4345 participants, began at noon on May 29, 2010, and ended 108 hours later (including hours during which the tournament was not running and the players presumably rested). Suppose we choose at random one of the 4344 players who were eliminated and let X denote the number of hours between this player's elimination and the previous elimination. If the chosen player was the first to be eliminated, then let X be the time between the start of the tournament and his elimination. What does the Markov inequality tell us about $P(X \geq 2.5)$?

Answer In total, 4344 eliminations occurred during 108 hours. Thus the mean time between eliminations was 108/4344 hours or $E(X)$ = 108/4344. By the Markov inequality, $P(X \geq 2.5) \leq$ 108/4344 ÷ 2.5 ~ 0.99%.

The Markov inequality is very crude. For instance, in Example 4.6.1, if $P(X \geq 2.5) = 0.99\%$, this would mean that all other 99.01% of the inter-elimination times were 0, and all of the inter-elimination times of at least 2.5 hours were exactly 2.5 hours.

A corollary of the Markov inequality is the **Chebyshev inequality,** which states that, for any random variable Y (not necessarily non-negative) with expected value μ and variance σ^2, for any real number $a > 0$,

$$P(|Y - \mu| \geq a) \leq \sigma^2/a^2.$$

Proof. Since $P(|Y - \mu| \geq a) = P[(Y - \mu)^2 \geq a^2]$, the result follows directly from the Markov inequality, letting $X = (Y - \mu)^2$ and $c = a^2$. ∎

Example 4.6.2

In the 2010 WSOP Main Event, each player began with 20,000 chips. Suppose after a few hours of play that the standard deviation of the number of chips among all the entrants is 8000. Using the Chebyshev inequality, find an upper bound for the probability that a randomly selected entrant has at least 60,000 chips.

Answer First, note that since the total number of chips in play does not change, the mean number of chips among all entrants remains 20,000. Second, note that if Y is the randomly selected player's number of chips, then since Y cannot possibly be less than 0, $Y - \mu$ cannot possibly be less than $-20,000$. Thus, $P(Y \geq 60,000) = P(Y - \mu \geq 40,000) = P(|Y - \mu| \geq 40,000)$, and according to Chebyshev's inequality, $P(|Y - \mu| \geq 40,000) \leq 8000^2/40,000^2 = 4.0\%$.

Incidentally, note that computation of this upper bound did not require knowledge of the number of players in the tournament or the number of hours played.

4.7 Moment-Generating Functions

The quantities $E(X)$, $E(X^2)$, $E(X^3)$, ... are called the *moments* of X. For any random variable X, its *moment-generating function* $\phi_X(t)$ is defined as $\phi_X(t) = E(e^{tX})$. Moment-generating functions have several useful properties. Their name comes from the fact that if one knows $\phi_X(t)$, then one can derive the moments $E(X^k)$ by taking derivatives of $\phi_X(t)$ and evaluating them at $t = 0$. Indeed, note that the derivative with respect to t of e^{tX} is Xe^{tX}, the second derivative is X^2e^{tX}, and so on. Thus, the kth derivative with respect to t of $\phi_X(t)$ is

$$(d/dt)^k E(e^{tX}) = E[(d/dt)^k e^{tX}] = E(X^k e^{tX}),$$

so

$$\phi_X'(0) = E\left(X^1 e^{0X}\right) = E(X), \phi_X''(0) \\ = E\left(X^2 e^{0X}\right) = E\left(X^2\right), \text{etc.} \tag{4.7.1}$$

(Note that in Equation 4.7.1, we are assuming that $(d/dt)\, E[e^{tX}] = E[(d/dt)\, e^{tX}]$, which technically requires justification but is a bit outside the scope of this book. The statement is true for all distributions considered here but may be invalid for continuous random variables with non-differentiable densities, for instance. See Billingsley [1990] for a treatment of this issue.)

Example 4.7.1

Consider again the random variable in Example 4.1.1, where $X = 1$ if you are dealt a pocket pair, and $X = 0$ otherwise. (1) What is $\phi_X(t)$? (2) What are $E(X^k)$ for $k = 1, 2, 3, \ldots$?

Answer

1. Recall that the probability of being dealt a pocket pair is $1/17$. Thus, $\phi_X(t) = E(e^{tX}) = e^{t(1)}$ $(1/17) + e^{t(0)} (16/17) = e^t/17 + 16/17$.
2. $\phi_X'(t) = e^t/17, \phi_X''(t) = e^t/17$, etc., so $\phi_X'(0) =$ $\phi_X''(0) = \ldots = 1/17$. Thus, $E(X^k) = 1/17$ for $k = 1, 2, \ldots$.

Example 4.7.2

For the random variable X in Example 4.1.2, what is $\phi_X(t)$? What are $E(X^k)$, for $k = 1, 2$, and 3?

Answer

$$\phi_X(t) = E(e^{tX}) = e^{1000t} (20\%) + e^{500t} (35\%) + e^{300t} (45\%).$$

$$\phi_X'(t) = 20\% \times 1000e^{1000t} + 35\% \times 500e^{500t} + 45\% \times 300e^{300t}$$

$$= 200e^{1000t} + 175e^{500t} + 135e^{300t},$$

so $E(X) = \phi_X'(0) = 200 + 175 + 135 = 510$, as we saw in Example 4.2.2.

$$\phi_X''(t) = 200 \times 1000e^{1000t} + 175 \times 500e^{500t} + 135 \times 300e^{300t},$$

so $E(X^2) = \phi_X''(0) = 200,000 + 87,500 + 40,5000$
$$= 328,000.$$

$$\phi_X'''(t) = 200,000 \times 1000 \; e^{1000t} + 87,500 \times 500 \; e^{500t}$$
$$+ 40,5000 \times 300 e^{300t},$$

so $E(X^3) = \phi_X'''(0) = 200,000,000 + 43,750,000 +$
12,150,000 = 255,900,000.

An important property of the moment-generating function that we use in Chapter 7 is that it uniquely characterizes the distribution of a random variable. Furthermore, for a sequence of random variables $X_1, \ldots, X_n$ with cdfs $F_i(y)$ and moment-generating functions $\phi_{Xi}(t)$, if $\phi_{Xi}(t)$ converges to some function $\phi(t)$, where $\phi(t)$ is the moment-generating function of X and F is the cdf of X, then $F_i(y) \to F(y)$ for all y where $F(y)$ is continuous. This is true whether the X_i values are discrete or continuous random variables. For proofs of these statements, see Billingsley (1990) or Feller (1967). The conclusion that $F_i(y) \to F(y)$ for all y where $F(y)$ is continuous is the definition of *convergence in distribution*, i.e., the equivalent of saying the distribution of X_t converges to the distribution of X.

Exercises

4.1 In their excellent book titled *Harrington on Hold'em*, Volume 1, Dan Harrington and Bill Robertie (2004) claim that "with a hand like AA, you really want to be one-on-one." Is this true even if you are all-in? To study this, do the following. In each scenario, suppose you are all-in before the flop with A♣ A♠, that you began the hand with $100 in chips, and that the blinds and/or antes are negligible.

a. Select a hand that you imagine A♣ A♠ might plausibly be heads up against, and using a poker odds calculator such as the one at www. cardplayer.com/poker_odds/texas_holdem, find the probability of your winning the hand and therefore ending up with $200 in chips. Then compute the expected value of your chips after the hand.

b. Select two hands you imagine A♣ A♠ might plausibly be up against and, as in part (a), compute the expected value of your chips after the hand.

c. Select three hands you imagine A♣ A♠ might plausibly be up against and again compute the expected value of your chips after the hand.

d. What do you conclude? Is it better to be all-in with AA against one opponent or several opponents?

4.2 Repeat Exercise 4.1, but substitute a low pair, like 55, instead of AA. Is it better to be all-in with 55 against one opponent or several opponents?

4.3 Suppose you have K♣ K♠, you go all-in for $100 in chips, and you are called by an opponent with 10♠ 10♦. Another player who has A♦ J♥ is considering whether to call also. In terms of maximizing your expected number of chips after the hand, do you want the other player to call? (Assume these are your only two opponents who might call, that they both have more than $100, and that the blinds and antes are negligible.) Using the poker odds calculator at www.cardplayer.com/poker_odds/texas_holdem to find the relevant probabilities, determine your expected number of chips after the hand (a) assuming the player with AJ folds and (b) assuming the player with AJ calls.

4.4 Suppose you play Texas Hold'em against nine opponents, in a casino where $5 is removed from every pot as a rake for the casino. Suppose 20 hands are dealt per hour. Let X be the average profit of the 10 players as a group after 3 hours. Can X be uniquely determined from this information alone? What is $E(X)$? What is $SD(X)$?

4.5 Consider the *High Stakes Poker* hand in Example 4.3.2, in which Minh Ly raised to $11,000 before the flop with K♥ K♦, Daniel Negreanu called with A♠ J♠, the flop came 8♠ 7♥ 2♠, and Ly bet $15,000. Although in the actual hand both players had far more chips, suppose for this problem that Ly's $15,000 on the flop was an all-in bet.

   a. If Negreanu calls, what is the probability that he will make a flush?

   b. If Negreanu calls, what is the probability that he will make a flush and lose the hand?

   c. If Negreanu calls, what is the probability that he will make a straight but not a flush?

   d. If Negreanu calls, what is the probability that he will not make a straight or a flush but win the hand anyway?

   e. Combine your answers from parts (a) through (d) to answer the following: If Negreanu calls, what is the probability that he will win the hand?

   f. Assuming Negreanu knew exactly what Ly's hole cards were, should he have called? Justify your answer. Assume the blinds and antes in this hand were negligible compared to the sizes of the other bets.

4.6 In a cash game, suppose you have 8♥ 7♥ and your opponent has K♥ K♦. You are heads-up, and the pot size is $200. The flop comes Q♥ 3♥ 2♣. Your opponent goes all-in, betting an additional $300. You have more than $300 left, so it is $300 more for you to call.

   a. If you call, what is the probability that you will make a flush?

   b. If you call, what is the probability that you will make a flush and your opponent will make a full house?

   c. If you call, what is the probability that you will make a flush and your opponent will make a higher flush?

   d. If you call, what is the probability that you will not make a flush but win the hand anyway?

e. Combine your answers from parts (a) through (d) to answer the following: If you call, what is the probability that you will win the hand?

f. Assuming you know exactly what cards your opponent has, should you call? Justify your answer.

4.7 *Phil Gordon's Little Blue Book* (p. 211) describes a hand from a no-limit Hold'em game at Caesar's Palace in Las Vegas. Gordon has A♣ 7♣, his opponent has 99, and the flop is 7♦ 7♥ 4♣ when both players get all-in. The pot size is $190,500.

a. Based on this information, using the poker odds calculator at www.cardplayer.com/poker_odds/ texas_holdem, find the probability of Gordon winning the hand.

b. Using probability, calculate the exact probability of Gordon winning the hand.

c. Gordon has more chips than his opponent at the beginning of the hand, so if Gordon wins the hand, his opponent will have $0 left. Let X denote the amount that Gordon's opponent will have after the hand is over. Using your answer in (b), calculate the expected value of X.

4.8 Suppose your opponent reraises all-in before the flop, and you know that she would do this with 90% probability if she had AA, KK, or QQ. If she had any suited connectors, she would do this with 20% probability. With any other hand, the probability that she would reraise all-in is 0. Given that she reraises all-in, what is the probability that she has suited connectors?

4.9 In a hand of the 2005 $1 million Bay 101 Shooting Star World Poker Tour event with three players left and blinds of $20,000 and $40,000 plus $5000 antes, the average chip stack was $1.4 million. The first player to act was Gus Hansen, who raised to $110,000 with K♦ 9♣. The small blind was Jay Martens, who reraised to $310,000 with A♣ Q♥.

The big blind folded and Hansen called. The flop was 4♦ 9♥ 6♣; Martens checked; Hansen went all-in for $800,000; and now Martens had a tough decision. He decided to call, at which point one of the announcers, Vince van Patten, said, "The doctor making the wrong move at this point. He still can get lucky of course." Was it the wrong move? Comment.

4.10 Consider the WSOP hand from the beginning of Chapter 1. Do you think Paul Wasicka should have called, given what he knew at the time? Explain your answer.

4.11 Find a hand involving only two players online or on TV and break it down according to luck and skill as in Examples 4.4.1, 4.4.2, or 4.4.3.

4.12 On one Hold'em hand in the 2009 WSOP $50,000 H.O.R.S.E. event, the summary at cardplayer.com described the action involving Jennifer Harman as follows: "Harman raised to $5,000 from under the gun and action folded to Max Pescatori in the cutoff. He three-bet to $7,500 and she was the only caller. The flop came down Q♠ 5♣ 4♦. She checked and Pescatori bet $2,500. Harman raised to $5,000 and Pescatori made the call. The turn was the J♣ and she led out with $5,000. Pescatori raised to $10,000 and she elected to muck her hand, leaving her with just $2,000 in chips." What probability of winning the hand would Harman have needed to call the raise on the turn if she wanted to maximize her expected number of chips? (Ignore blinds and antes in your calculation.) What do you think of her fold on the turn? What would you guess she and Pescatori had?

4.13 On one hand during Episode 2 of Season 7 of *High Stakes Poker*, with blinds of $400 and $800 plus $100 antes from eight players, Antonio Esfandiari raised to $2500 with 8♥ 7♥. Four players called, including the small blind and Barry Greenstein, who had 4♦ 4♣ in the big blind. The flop came 10♣ 6♥ 4♥. Greenstein checked; Esfandiari bet $6200; David Peat called; Greenstein raised to $30,000; Esfandiari reraised to $106,000; Peat folded; and Greenstein went all-in for an additional $181,200. If Esfandiari *knew* what Greenstein had at this point, should he have called? Explain your answer. (In reality, Esfandiari called. The turn was Q♥ and the river 8♣, so Esfandiari won the $593,900 pot.)

4.14 Consider the situation described in Example 3.4.3. Using the assumptions about the distribution of Phil Hellmuth's possible hole cards in the example, should Phil Gordon have called or folded? Note: unlike the solution to Example 3.4.3b, you should take the possibility of a tie into account in your answer here.

4.15 Suppose that X has moment-generating function $\phi_X(t)$ and let $Y = 3X + 7$. (a) Find the moment-generating function of Y. (b) Suppose $\phi_X(6) = 0.001$. What is $\phi_Y(2)$ where $\phi_Y(t)$ is the moment-generating function of Y?

4.16 After winning a six-handed, winner-take-all tournament on *Poker After Dark* in 2009, Vanessa Rousso claimed she won by "calculating probabilities and thinking about pot odds." In one key hand when the tournament was three-handed, with blinds of $800 and $1600, David Grey raised to $4000 on the button, Jennifer Harman folded in the small blind, and Rousso called with 10♥ 7♥. The pot then was $8800 in chips. After the flop of K♥ 4♠ Q♥ was revealed, Rousso checked, Grey bet $5000, Rousso raised to $15,000, Grey went all-in for $44,100 total, and Rousso thought about her decision. The pot was $67,900, and it cost $29,100 more for Rousso to call. Rousso said, "I could be drawing dead. I'm gonna fold," and she showed her hand and folded. Was this a smart fold? Explain your answer. (As it turned out, Grey had K♠ 10♠.)

4.17 Suppose as in Examples 4.3.5 to 4.3.7 that you are heads up on the river and you cannot possibly win the pot in a showdown. You are considering either checking or making a bet of half the size of the pot. With what probability would your bluff need to be successful in order for it to be profitable in the long term?

4.18 Suppose you are heads up on the river, your opponent bets the size of the pot, and you are considering calling or folding. With at least what probability does calling have to be correct in order for calling to be more profitable in the long term?

4.19 Suppose you are heads up on the river, your opponent bets half the size of the pot, and you are considering calling or folding. With at least what probability does calling have to be correct in order for a call to be more profitable in the long term?

4.20 On day 4 of the 2015 WSOP Main Event, with blinds of 4000 and 8000 plus antes of 1000 from each of the eight players, the first six players folded and the pot was still 20,000 when the small blind Moreira De Melo with J♥ 9♣ raised to 17,000, and the big blind, Ka Kwon Lau, called 9,000 with 5♦ 3♥. The pot was 42,000. The flop came 10♠ K♥ 5♥. De Melo bet 18,000 and Lau called. The pot was now 78,000. The turn was K♣. De Melo checked, Lau bet 21,000, and De Melo called. The pot was 120,000. The river was Q♣ and they both checked. The announcer Norman Chad on the ESPN broadcast said of De Melo's play, "I can't say I understand this check call on the turn." (a) Given the cards De Melo and her opponent had and the four cards on the board, was De Melo's call on the turn correct in terms of maximizing her expected number of chips? (b) How much expected profit did Lau gain due to skill on the turn? (c) How much did De Melo gain due to luck on the river?

4.21 On day 5 of the WSOP Main Event in 2015, Chad Power raised to 50,000 in early position with J♥

10♦, and the big blind, Sal DiCarlo, called with A♣ 8♣. Because of blinds and antes, the pot was now 148,000. The flop was 9♣ K♣ 4♦. DiCarlo checked, Power bet 40,000, and DiCarlo called. After the Q♥ was dealt on the turn, DiCarlo checked, Power bet 115,000, and DiCarlo called, making the pot 458,000. The river was 6♥, DiCarlo checked, Power bet 500,000, and DiCarlo folded. ESPN announcer Lon McEachern said "Chad Power will take that pot. Very nicely played by the poker teacher." However, did he win more by luck or skill? Using a poker odds calculator such as the one at cardplayer.com to obtain the relevant probabilities, calculate how much expected profit Power won (a) due to luck on the flop, (b) due to skill on the flop, (c) due to luck on the turn, (d) due to skill on the turn, (e) due to luck on the river, and (f) due to skill on the river. (g) Summing up your answers, how much in total expected profit did Power win on the flop, turn, and river due to luck and how much due to skill?

4.22 At the 2015 WSOP Main Event, with just two players left in the tournament, Josh Beckley raised with Q♣ 7♠ and Joe McKeehen called with J♥ 8♥. The board came J♠ J♣ 6♠ 3♣ 9♣. Josh Beckley bluffed with queen high on the flop and turn, and McKeehen called each time. On the river, the pot was 14.5 million chips before Beckley bluffed 4.7 million. How often would such a bluff have to work in order to give Beckley positive expected value? (McKeehen called before collecting the 23.9 million chip pot and ultimately winning the event, becoming the world champion and earning nearly $7.7 million for first place.)

4.23 The term *semi-bluff* is used to describe situations where the player betting is currently behind but has outs and thus can have a decent chance to win even if the opponent calls. Semi-bluffs can be highly profitable.

In a hand between Daniel Negreanu and Tom Cannuli on day 6 of the 2015 WSOP Main Event, Negreanu raised to 115,000, and Tom Cannuli called with 7♠ 6♠. With blinds and antes, the pot was 345,000. The flop came J♠ K♥ 9♠, Negreanu checked, and Cannuli semi-bluffed 160,000. Suppose that, given what Negreanu might have, there is a 50% chance Negreanu will fold to the semi-bluff, and 50% chance he will call. Assuming no further betting will occur, what is the expected profit Cannuli obtains by semi-bluffing, compared to checking? (In the actual hand, Negreanu had 8♠ 8♦ and quickly folded.)

4.24 Sometimes semi-bluffs (see previous exercise) do not work out. On day 6 of the 2015 WSOP Main Event, physician Wasim Ahmar raised on the button with Q♥ 7♥. Daniel Negreanu called with A♣ 2♣ and Tom Cannuli called as well. The pot was 840,000. The flop came 9♥ K♠ A♥. Negreanu and Cannuli checked, Ahmar semi-bluffed 450,000, Negreanu called, and Cannuli folded. The pot was now 1.74 million. The turn was Q♦. Negreanu bet just 350,000 and Ahmar folded! (a) After Negreanu bet on the turn, what probability did Ahmar need to win in order for a call to give him a higher expected number of chips than folding? (b) Given the four cards on the board and the cards Negreanu and Ahmar had, what was Ahmar's probability of winning the hand on the river assuming no further betting? (c) In the actual hand, Ahmar guessed Negreanu had J10 before folding. In the theoretical worst possible case where Negreanu had J♥ 10♥, what would Ahmar's probability of winning the hand have been, assuming no further betting?

4.25 Sometimes both players are semi-bluffing (or calling with draws) and on the river neither has anything in the end. An example is in the 2015 WSOP Main Event on day 6, where Neil Blumenfield had Q♦ J♠, Wasim Ahmar had A♠ 2♠, and the board came 9♠ 10♠ 6♣ 5♣ 6♥ before Blumenfield bluffed all-in and Ahmar folded. Given the hands Blumenfield and Ahmar had, what is the probability that the board would completely miss both players' hands? Specifically, ignore the possibility of straights or flushes and simply calculate the probability that none of the five board cards is a Q, J, A, or 2, given the cards Blumenfield and Ahmar had.

4.26 Being tricky can be highly profitable, but giving away free cards can be dangerous. With just 22 players left in the 2015 WSOP Main Event, after an initial raise by John Allen Hinds, Neil Blumenfield reraised with A♠ A♥ and Hinds called with A♦ Q♣. The pot was 2.47 million when the flop came 8♥ 4♣ 10♣. Blumenfield bet just 450,000 and Hinds called. The turn was A♣, Hinds checked, and Blumenfield, perhaps trying to be tricky, checked also. However, the river was 7♣, Hinds bet 1.7 million, Blumenfield called, and Hinds took down the 6.77 million chip pot. Calculate how much expected profit Blumenfield lost due to luck on the river, and compare to how much he lost due to skill on the river.

CHAPTER 5

Discrete Random Variables

M ost poker problems involve discrete random variables, since the number of cards in the deck and therefore the number of combinations of possible outcomes are always finite. Similarly, the hands played by a person or even those played by an entire population are countable and must be finite over any finite time span, so even when considering collections of hands, discrete random variables dominate most applications.

Some special examples of discrete random variables arise so frequently that they have names and deserve special attention. Among these are Bernoulli, binomial, geometric, Poisson, and negative binomial random variables, and all arise naturally in analyzing the outcomes of hands of Texas Hold'em.

We will follow the usual probabilistic terminology of describing these variables in terms of the results of independent *trials*. The word *trial* is meant to be general and can refer to all sorts of different things, for example, the recorded speed of an animal or other experimental subject studied under controlled laboratory conditions or the result of a question asked of a person sampled at random from a large population. In our examples, a *trial* will typically refer to a hand of Texas Hold'em or a tournament. It seems reasonable to suppose that the results of such trials will often be close to independent.

5.1 Bernoulli Random Variables

If a random variable X can only take the values 0 or 1, then we say X is a *Bernoulli* random variable. More precisely, we say X is *Bernoulli*(p) if $X = 1$ with probability p and $X = 0$ with probability q, where $q = 1 - p$. The pmf of X is $f(1) = p, f(0) = q$, and $f(b) = 0$ for all other values of b.

One example of a Bernoulli random variable was given in Example 4.1.1. These two-valued random variables are named after Jacob Bernoulli (1654–1705), a Swiss mathematician whose most famous work, *Ars Conjectandi*, applied probability to games of chance and introduced the law of large numbers.

The convention is for Bernoulli random variables to take the values 0 and 1, rather than 1 and –1, for good reason. With the given convention, the *sum* of n Bernoulli random variables represents the total number of successes (1s) among n trials. The *average* of Bernoulli random variables is the percentage of successes, which is an unbiased estimate of p.

Example 5.1.1

Suppose you and nine other players play in a weekly tournament, and you are all equal in ability. Let $X = 1$ if you win, and 0 if you lose. What is the pmf of X? What is the cdf of X?

Answer By assumption, the 10 participants all play equally well, so $p = 1/10$. Thus the pmf of X is $f(1) = 1/10, f(0) = 9/10$, and $f(b) = 0$ for $b \neq 0$ or 1. The pmf is plotted in Figure 5.1.

The cdf of X (Figure 5.2) is

$$F(b) = 0 \text{ for } b < 0,$$

$$F(b) = 9/10 \text{ for } 0 \leq b < 1,$$

$$F(b) = 1 \text{ for } 1 \leq b.$$

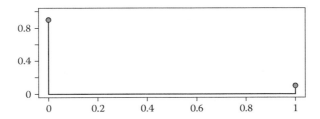

FIGURE 5.1 Probability mass function of a Bernoulli (0.10) random variable.

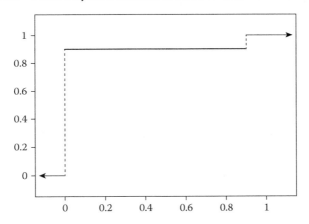

FIGURE 5.2 Cumulative distribution function for a Bernoulli (0.10) random variable.

For a Bernoulli random variable X, the expected value $E(X) = p$ and the standard deviation $\sigma = \sqrt{(pq)}$. Indeed, referring to the formulas for expected value and variance,

$$E(X) = (1 \times p) + (0 \times q) = p,$$

and

$$\begin{aligned} \mathrm{var}(X) &= \left[E(X^2) \right] - (E(X))^2 \\ &= \left[(1^2 \times p) + (0^2 \times q) \right] - p^2 \\ &= p - p^2 \\ &= pq. \end{aligned}$$

Example 5.1.2

Suppose $X = 1$ if you are dealt a pair on your next hand, and otherwise $X = 0$. What are the expected value and standard deviation of X?

Answer The probability that you will have a pocket pair is $13 \times C(4,2)/C(52,2) = 1/17$ ~ 5.88%, so $E(X) = p$ ~ 0.0588, and the standard deviation is

$$\sigma = \sqrt{(pq)} = \sqrt{(1/17 \times 16/17)} \sim 0.235.$$

5.2 Binomial Random Variables

Suppose X represents the number of times an event occurs out of n independent trials and the event has probability p in each trial. Then X is a binomial random variable and its short form is $X \sim binomial\,(n,p)$. In this case, X can take any of the $n + 1$ possible integers in the set $\{0,1,2,\ldots,n\}$. If k is an integer in this set, then there are $C(n,k)$ possible collections of outcomes of the n trials so that the event occurs in exactly k of them. The probability associated with each such collection of outcomes with the event occurring k times and the event not occurring the other $n - k$ times is simply $p^k\,q^{n-k}$, where $q = 1 - p$, because the trials are independent. Thus the pmf of X is $f(k) = C(n,k)\,p^k\,q^{n-k}$. The following example may help to illustrate this in more detail.

Example 5.2.1

Suppose X is the number of pocket pairs you are dealt in the next seven hands. Derive $P(X = 3)$.

Answer Letting 1 denote a pair and 0 denote a non-pair, we might write $\{1,1,1,0,0,0,0\}$ to represent the event that you are dealt pairs on your first three hands and non-pairs on the next four hands.

The probability associated with this exact collection of outcomes is clearly p^3q^4, where p is the probability of being dealt a pair = 1/17, and $q = 1 - p$ = 16/17. However, there are other ways you could get exactly three pairs in the next seven hands: the outcomes could be {1,0,1,1,0,0,0} or {0,0,1,1,1,0,0}, etc. Each of these possible collections has probability p^3q^4, and there are $C(7,3) = 35$ such collections because there are $C(7,3)$ different ways of choosing exactly three of the seven trials to be dealt a pocket pair.

Thus $P(X = 3) = C(7,3)\,(1/17)^3\,(16/17)^4$.

The abbreviation *iid* means *independent and identically distributed*. The independence concept was discussed in Chapter 3. *Identically distributed* means that the probability p for some event in question does not change from trial to trial. In the definition of binomial random variables, it is important that the trials are *iid*. If the trials are not independent, or if the probability of the event differs for different trials, then the number of occurrences of the event in n trials may have a distribution very different from the binomial.

Note that the sum of independent Bernoulli random variables is a binomial random variable. That is, if Y_1, $Y_2, \ldots, Y_n$ are independent *Bernoulli*(p) random variables, and $X = Y_1 + Y_2 + \ldots + Y_n$, then $X \sim binomial(n,p)$. This is another advantage of defining Bernoulli random variables as {0,1} valued. As we will see in Section 7.1, the mean of the sum of independent random variables is equal to the sum of the means, and similarly the variance of the sum is equal to the sum of the variances. As a result, the mean and variance of the binomial random variable are simply n times those of the Bernoulli random variable.

If $X \sim binomial(n,p)$, then $E(X) = np$, and var$(X) = npq$.

Higher moments, i.e., $E(X^k)$ for $k \geq 2$, can be obtained via the moment-generating function. If X is *binomial(n,p)*, then its moment-generating function $\phi_X(t) = E\left(e^{tX}\right) = \sum_{k=0}^{n} e^{tk} C(n,k) p^k q^{n-k} = \sum_{k=0}^{n} C(n,k) \left(pe^t\right)^k q^{n-k} = \left(pe^t + q\right)^n.$

Example 5.2.2: Playing the Board

On the last hand of the 1998 WSOP Main Event, with the board 8♣ 9♦ 9♥ 8♥ 8♠, Scotty Nguyen went all-in. While his opponent, Kevin McBride, was thinking, Scotty said, "You call, it's gonna be all over, baby." McBride said, "I call. I play the board." It turned out that Scotty had J♦ 9♣ and won the hand. Assuming you never fold in the next 100 hands, what would be the expected value of X = the number of times in these 100 hands that you would *play the board* after all five board cards are dealt? What is the standard deviation of X?

Answer Since we may assume that what occurs on each hand is independent of the other hands, $X \sim binomial(n,p)$, where $n = 100$ and p = the probability of playing the board. Each collection of seven cards consisting of your two hole cards and the five cards on the board is equally likely to occur. For each such collection of seven cards, there are $C(7,2)$ different ways of choosing two of the cards to be your two hole cards. Each such choice is equally likely and only one of these choices results in your two hole cards being the two cards that do not *play*, so $p = 1/C(7,2) = 1/21$. Thus $E(X) = 100 \times (1/21) \sim 4.76$. $\text{var}(X) = npq = 100 \times (1/21) \times (20/21) \sim 4.54$, and $SD(X) = \sqrt{(npq)} \sim 2.13$.

(Note that this calculation is a slight underestimation of the probability of playing the board because, in rare cases, it is ambiguous whether you are using your hole card or the board in forming your best five-card hand, e.g., when you have JJ and the board is 2222J or AKQ J10. In these situations you are generally considered to be playing the board, but we did not count them correctly as such in the solution to Example 5.2.2.)

5.3 Geometric Random Variables

Suppose as in Section 5.2 that you observe the results of repeated iid trials and that X is the number of trials until the *first* occurrence of some event that has a constant probability p of occurring on each trial. Then X is a geometric random variable, and we write $X \sim geometric(p)$. For example, X may be the number of hands until you are dealt pocket aces, as in Example 5.3.1.

By convention, the counting includes the hand that deals you pocket aces, that is, if you get AA on your first hand, then $X = 1$. Thus, a geometric random variable may take values in the positive integers only. The pmf of a *geometric(p)* random variable is simply $q^{k-1}p$, for $k = 1,2,\ldots$, where $q = 1 - p$, because in order for X to equal k, the first $k - 1$ trials must all *not* contain the event, and then the kth trial must contain the event, and by assumption the outcomes of the trials are independent. The cdf of X is $F(k) = P(X \le k) = 1 - P(X > k) = 1 - q^k$, for $k = 0,1,2,\ldots$, because X exceeds k if and only if the event does not occur in the first k trials.

If X is *geometric(p)* with $p > 0$, then $E(X) = 1/p$ and $\text{var}(X) = q/(p^2)$.

To prove that $E(X) = 1/p$, the following lemma is useful. It applies not only to geometric random variables but also to any random variable taking on only nonnegative integer values.

Lemma 5.3.1

Suppose X is a random variable taking only values in $\{0,1,2,3, \ldots\}$. Then $E(X) = \sum_{k=0}^{\infty} P(X > k)$.

Proof.

$$\sum_{k=0}^{\infty} P(X > k) = P(X > 0) + P(X > 1) + P(X > 2) + \dots$$

$$
\begin{aligned}
&= P(X = 1) + P(X = 2) + P(X = 3) + P(X = 4) + \dots \\
&\quad + P(X = 2) + P(X = 3) + P(X = 4) + \dots \\
&\quad + P(X = 3) + P(X = 4) + \dots
\end{aligned}
$$

$$= \sum_{k=0}^{\infty} kP(X = k)$$

$$= E(X). \qquad ■$$

The function $P(X > k)$ is called the *survivor function*. Lemma 5.3.1 describes how one may write the expected value of a non-negative integer-valued random variable in terms of the survivor function. For geometric random variables, the survivor function has the particularly simple form of $P(X > k) = q^k$, for $k = 0,1,2,\dots$. To find the expected value of a *geometric(p)* random variable with $p > 0$, observe that, using Lemma 5.3.1,

$$E(X) = \sum_{k=0}^{\infty} P(X > k)$$

$$= \sum_{k=0}^{\infty} q^k = 1/(1-q) = 1/(1-(1-p)) = 1/p.$$

Obviously, if $p = 0$, then X is always infinite, so $E(X) = \infty$.

Example 5.3.1

On *High Stakes Poker* Season 3, Paul Wasicka folded many hands in a row after first joining the table, prompting Mike Matusow to joke, "Don't worry, Paul, those aces are coming soon." Of course, Wasicka played many other hands, but suppose you were truly waiting for pocket aces. How many hands would you expect to have to wait and what would be the standard deviation of the number of hands to wait?

Answer Let X = the number of hands to wait before getting aces. $X = geometric(p)$, where p = the probability of being dealt AA on a particular hand, which is $C(4,2) \div C(52,2) = 1/221$. Thus $E(X) = 1/p = 221$, and

$$var(X) = q/p^2 = 220/221 \div (1/221)^2 = 48620,$$

$$\text{so } SD(X) = \sqrt{48620} \sim 220.5.$$

Example 5.3.2

Until he finally won a WSOP event in 2008, Erick Lindgren was often called one of the greatest players never to have won a WSOP tournament. Before his win, he played in many WSOP events and finished in the top 10 eight times. Suppose you play in one tournament per week. For simplicity, assume that each tournament's results are independent of the others and that you have the same probability p of winning each tournament. If $p = 0.01$, then what is the expected amount of time before you win your first tournament, and what is the standard deviation of this waiting time?

Answer The waiting time $X \sim geometric(0.01)$, so $E(X) = 1/0.01 = 100$ weeks, and $SD(X) = (\sqrt{0.99})/0.01 \sim 99.5$ weeks.

If $X \sim geometric(p)$, then *conditional on the fact that $X > c$*, for some positive integer c, the pmf of X is simply $f(k - c)$, where f is the unconditional pmf of X (see Exercise 5.3). In other words, knowledge that the event in question has not occurred in the first c trials does not affect the number of *additional* trials needed for the event to occur. Since the trials are independent, it makes sense that information on past trials would not influence the probability p governing the likelihood of the event on future trials. The geometric distribution is the only discrete distribution that has this *memorylessness* property.

5.4 Negative Binomial Random Variables

Recall from Section 5.3 that the number of iid trials required before the *first* time a certain event occurs is a geometric random variable. However, if you are interested in the number of such trials until the rth occurrence of the event, where r is a positive integer, then X is a *negative binomial* (r,p) random variable. An example is the number of hands until your third pocket pair, in which case X can be any value in the set $\{3,4,5,\ldots,\infty\}$.

For the rth event to occur on exactly the kth trial, there must be exactly $r-1$ occurrences of the event on the first $k-1$ trials, and then the event must occur on the kth trial. For instance, in order to get the third pocket pair on hand 10, you must have had two pocket pairs by hand 9, and then subsequently get a pocket pair on hand 10. Hence we can see that the pmf for a negative binomial (r,p) random variable X is given by $f(k) = C(k-1, r-1)\, p^r\, q^{k-r}$, for $k = r, r+1, \ldots$, where $q = 1-p$.

For a negative binomial (r,p) random variable X, the moment-generating function $\phi_X(t) = [pe^t/\{1 - qe^t\}]^r$. From this (or directly, using the pmf) one can obtain $E(X) = r/p$ and $\text{var}(X) = rq/p^2$.

Example 5.4.1

During Episode 2 of Season 5 of *High Stakes Poker*, Doyle Brunson was dealt pocket kings twice and pocket jacks once, all within about half an hour. Suppose we consider a *high pocket pair* to mean 10 10, JJ, QQ, KK, or AA. Let X be the number of hands you play until you are dealt a high pocket pair for the third time. What is $E(X)$? What is $SD(X)$? What is $P(X = 100)$?

Answer   The probability of being dealt a high pocket pair on a certain hand is $5 \times C(4,2)/C(52,2) = 5/221$, so

$$E(X) = 3 \div 5/221 = 132.6.$$

$$SD(X) = \sqrt{(rq)}/p$$

$$= \sqrt{(3 \times 216/221)}/(5/221)$$

$$\sim 75.7.$$

$$P(X = 100) = C(99,2)\,(5/221)^3\,(216/221)^{97} \sim 0.61\%.$$

5.5 Poisson Random Variables

On page 57 of *Harrington on Hold'em*, Volume 2, Harrington and Robertie (2005) suggest making a bluff about every hour and a half at a typical table against typical opponents. They also suggest randomizing such strategic decisions, for instance, by using a watch. Suppose three players attempt to take that advice. Player 1 plays in a very slow game (only about four hands an hour), and she decides to do a big bluff whenever the second hand on her watch, at the start of the deal, is in some predetermined 10-second interval. It seems reasonable to assume that the positions of the second hand on her watch are approximately independent from one poker hand to another. Thus, Player 1 may occasionally bluff a few hands in a row, but on average she is bluffing once every six hands, or once every 90 minutes. Suppose Player 2 plays in a game in which about 10 hands are dealt per hour. He also looks at his watch at the beginning of each poker hand, but only does a big bluff if the second hand is in a 4-second interval. Player 3 plays in a faster game (about 20 hands dealt per hour). She bluffs only when the second hand on her watch at the start of the deal is in a 2-second interval. Each of the three players will thus average one bluff every hour and a half.

Let X_1, X_2, and X_3 denote the numbers of big bluffs attempted in a given 6-hour interval by Player 1, Player 2, and Player 3, respectively. Each of these random variables is binomial with an expected value of 4, and a variance approaching 4. However, there are some noticeable differences between the three distributions. Player 1 will only have played 24 hands in a typical 6-hour interval, in which case the highest possible value of X_1 would be 24. If Players 2 and 3 play 60 and 120 hands in the same period, then X_2 could be any integer between 0 and 60, and X_3 any number from 0 to 120, although values higher than 24 would be exceedingly rare.

Figure 5.3 shows the probability mass functions between 0 and 20 for each of the three random variables. One can see that they converge toward some limiting distribution known as the *Poisson distribution*. Unlike the binomial distribution that depends on two parameters, n and p, the Poisson distribution depends only on one parameter, λ, called the *rate*. In our example, $\lambda = 4$.

The pmf of the Poisson random variable is $f(k) = e^{-\lambda} \lambda^k / k!$, for $k = 0,1,2,\ldots$, and for $\lambda > 0$, with the convention that $0! = 1$, and where $e = 2.71828 \ldots$ is Euler's number. As discussed above, the Poisson random variable is the limit in distribution of the binomial distribution as $n \to \infty$ while np is held constant at the value λ. To prove this, note that the pmf for a binomial random variable is

$$f(k) = C(n,k)\, p^k\, q^{n-k}$$

$$= C(n,k)\, (\lambda/n)^k\, (1 - \lambda/n)^{n-k}$$

$$= C(n,k)/n^k\, \lambda^k\, (1 - \lambda/n)^{-k}\, (1 - \lambda/n)^n.$$

Thus, using the calculus identity $\lim_{n\to\infty} (1 - \lambda/n)^n = e^{-\lambda}$, for $\lambda > 0$, and observing that for any fixed non-negative integer k, $\lim_{n\to\infty} C(n,k)/n^k = \lim_{n\to\infty} n!/[k!\,(n-k)!\,n^k]$

$$= 1/k!\, \lim_{n\to\infty} n(n-1)(n-2)\ldots(n-k+1)/n^k$$

$$= 1/k!\, \lim_{n\to\infty} [n/n]\,[(n-1)/n]\,[(n-2)/n]\ldots[(n-k+1)/n]$$

$$= 1/k!,$$

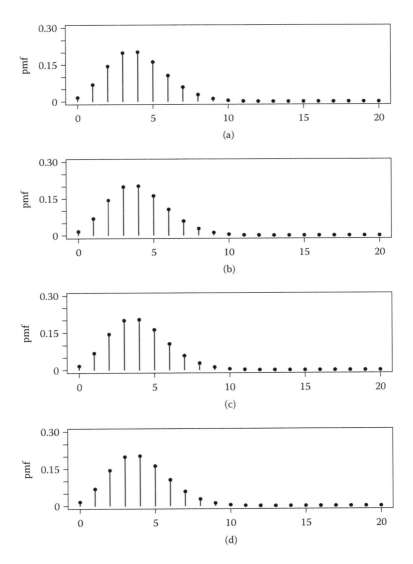

FIGURE 5.3 Probability mass functions for binomial random variables with mean 4: (a) binomial (24, 1/6); (b) binomial (60, 1/15); (c) binomial (120, 1/30); and (d) Poisson (4).

and $\lim_{n\to\infty}(1 - \lambda/n)^{-k} = 1$, we have $\lim_{n\to\infty} f(k) = 1/k!\ \lambda^k\ e^{-\lambda}$. If X is a *Poisson*(λ) random variable, then its moment-generating function

$$\phi_X(t) = E\left(e^{tX}\right)$$

$$= \sum_{k=0}^{\infty} e^{tk} e^{-\lambda}\ \lambda^k/k!$$

$$= e^{-\lambda} \sum_{k=0}^{\infty} \left(\lambda e^t\right)^k/k!$$

$$= \exp\{-\lambda\}\exp\{\lambda e^t\}$$

$$= \exp\left(\lambda e^t - \lambda\right).$$

Examining $\phi_X'(0)$ and $\phi_X''(0)$, one observes (see Exercise 5.8) that for a *Poisson*(λ) random variable X, $E(X) = \lambda$ and var(X) = λ also. Note that λ must be a positive real number but need not be an integer. In many examples, such as the example discussed in the beginning of this section of the number of bluffs in an hour, or Example 5.5.1, the parameter λ can be interpreted as a *rate*, indicating the mean number of occurrences of some event per unit of time, or per unit of the domain in which the events are observed.

In fact, the Poisson distribution also arises as the number of events occurring in a given time span, provided the numbers of events in any disjoint time spans are independent. Such collections of events are thus called *Poisson processes*. For instance, the occurrences of wildfires, hurricanes, disease epidemics, UFO sightings, and incidence of particular species have all been modeled as Poisson processes. By contrast, events that tend to occur in clusters, such as the locations and times of earthquakes, or events that are inhibitory and tend to have very few clusters, such as the locations of giant redwood trees, are generally not well modeled as Poisson processes.

Example 5.5.1

Many casinos award prizes for rare events called *jackpot hands* (see Example 3.3.3). These jackpot hands are defined differently by different casinos. Suppose in a certain casino jackpot hands are defined so that they tend to occur about once every 50,000 hands on average. If the casino deals about 10,000 hands per day, what are the expected value and standard deviation of the number of jackpot hands dealt in a 7-day period? How close are the answers using the binomial distribution and the Poisson approximation? Using the Poisson model, if X represents the number of jackpot hands dealt over the week, what is $P(X = 5)$? What is $P(X = 5 \mid X > 1)$?

Answer It is reasonable to assume that the outcomes of different hands are iid, and this applies to jackpot hands as well. In a 7-day period, approximately 70,000 hands are dealt, so X is binomial ($n = 70,000$, $p = 1/50,000$). Thus

$$E(X) = np = 1.4,$$

and $SD(X) = \sqrt{(npq)}$

$$= \sqrt{(70,000 \times 1/50,000 \times 49,999/50,000)}$$

$$\sim 1.183204.$$

Using the Poisson approximation,

$$E(X) = \lambda = np = 1.4,$$

and $SD(X) = \sqrt{\lambda} \sim 1.183216.$

The Poisson model is a very close approximation in this case. Using the Poisson model with rate
$\lambda = 1.4$,

$$P(X = 5) = \exp(-1.4)1.4^5/5!$$

$$\sim 0.01105, \text{ or about 1 in } 90.5.$$

$$P(X = 5 \mid X > 1) = P(X = 5 \text{ and } X > 1) \div P(X > 1)$$

$$= P(X = 5) \div P(X > 1)$$

$$= \left[\exp(-1.4)1.4^5/5! \right]$$

$$\div \left[1 - \exp(-1.4)1.4^0/0! \right.$$

$$\left. - \exp(-1.4)1.4^1/1! \right]$$

$$\sim 0.01105 \div 0.4082$$

$$= 0.0271, \text{ or about 1 in } 36.9.$$

Exercises

5.1 Doyle Brunson famously won the WSOP Main
Events in 1976 and 1977, each time holding the hole
cards (10, 2) on the final hand, and each time making a full house on the river. Out of $n = 100$ hands,
suppose X is the number where you are dealt (10, 2)
and make a full house by the time the river card is
dealt. Find $P(X \geq 2)$.

5.2 Suppose you repeatedly play hands of Texas
Hold'em, and let $X_1 =$ the number of hands you are
dealt until you get a pocket pair, and $X_2 =$ the number of hands until you are dealt two black cards. Let
$Y = min\{X_1, X_2\}$ and $Z = max\{X_1, X_2\}$. Find general
expressions for the pmf of Y and the pmf of Z.

5.3 Suppose $X \sim geometric(p)$, and let f denote the pmf
of X. Show that, if $k > c$ where k and c are integers,
then $P(X = k \mid X > c) = f(k - c)$, thus proving that for

a geometric random variable, the information that the event in question has not occurred in the first c trials does not affect the number of additional trials needed for the event to occur.

5.4 Suppose $X_1 \sim Poisson(\lambda_1)$ and $X_2 \sim Poisson(\lambda_2)$ and that X_1 and X_2 are independent. Using the result of Exercise 7.12 that $E(f(X_1)\,g(X_2)) = E(f(X_1))\,E(f(X_2))$, show that $Y = X_1 + X_2$ is also Poisson distributed, with mean $\lambda_1 + \lambda_2$.

5.5 Let X denote the number of hands until you are dealt pocket aces for the first time, and let Y denote the number of hands until you are dealt a high pocket pair for the first time, with *high* defined as in Example 5.4.1. Compare the expected value and standard deviation of X with the expected value and standard deviation of Y.

5.6 The 2008 WSOP Main Event final table lasted 15 hours and 28 minutes and consisted of 274 hands. On the televised broadcast, only 23 of the hands were shown, and among these, eventual champion Peter Eastgate played eight of the hands, won all eight and made three of a kind or better on six of the hands. (a) If you never fold, how often will you make three of a kind or better? (b) Assuming the table deals 274 hands over 15 hours and 28 minutes and the player in question never folds, what are the expected value and standard deviation of the number of times a player will make three of a kind or better during the 15 hours and 28 minutes? How close are the answers using the binomial distribution and the Poisson approximation? (c) Using the Poisson model, if X represents the number of three-of-a-kind or better hands a player gets over this 15 hour and 28 minute period, what is $P(X = 6)$? What is $P(X \geq 6)$?

5.7 Show that, for a geometric random variable X with parameter p, $var(X) = q/p^2$.

5.8 Using the derivatives of the moment-generating function, show that if X is a *Poisson*(λ) random variable, then $E(X) = \text{var}(X) = \lambda$.

5.9 Show that the moment-generating function for a negative binomial (r,p) random variable is $\phi_X(t) = [pe^t/\{1 - qe^t\}]^r$.

5.10 Show that a negative binomial (r,p) random variable X has mean $E(X) = r/p$ and variance rq/p^2.

5.11 In the 2015 WSOP Main Event, Kelly Minkin was eliminated in 29th place and was the last woman remaining in the tournament. About 4% of the 6420 entrants were women. Suppose all players play identically, so there is a probability of 4% that the first place player is female, 4% probability that the second place player is female, etc., and assume for the purpose of this problem that the genders of different finishers are independent. (In reality they would be nearly independent, but not quite independent.) Let X = the place where the last remaining woman in the draw gets eliminated (or $X = 1$ if a woman wins). (a) What type of random variable is X? (b) What are $E(X)$ and $SD(X)$? (c) cardplayer.com lists the places where the last remaining women exited the WSOP Main Event each year since 2004 as 98, 15, 56, 38, 17, 27, 121, 29, 10, 31, 77, and 29 in the year 2015. Do these results seem consistent with your answer to part (b)? There is no need to do any calculation in this part of the problem. Just see, by eye, if the results appear roughly consistent with your answer in (b).

5.12 In 2014, Ronnie Bardah accomplished the amazing task of cashing (earning prize money) in his fifth consecutive WSOP Main Event. Suppose the tournament has the same 6000 entrants per year, 10% of whom cash, every player has the same chance of cashing approximately independently of any other player, and results in different years are independent events. (a) If you play five times in the Main

Event, what is the probability that you will cash all five times? (b) If X is the number of years until *someone* cashes five times in a row again, what type of random variable is X? (c) What are $E(X)$ and $SD(X)$?

5.13 In the 2015 WSOP Main Event, only 82 of the 6420 entrants were German. Suppose each of the 6420 entrants had the same chance of finishing in the top 25 and that whether one player finished in the top 25 was approximately independent of whether any other player did the same. Let X denote the number of Germans making the final 25. What type of random variable is X? What are $E(X)$ and $SD(X)$? (It turned out that, remarkably, there were 4 Germans of the final 25 competitors.)

CHAPTER 6

Continuous Random Variables

Nearly all probability problems surrounding Texas
Hold'em involve discrete and typically finite possibilities, as there are only 52 cards in the deck and thus
only a finite number of possible outcomes on a given hand,
and since hands are clearly separated so one can only possibly play an integral number of hands. It does not make
sense to talk about playing 2.5 hands or e hands, for example. However, in Section 5.5, we saw that limiting distributions can be helpful, as for instance Poisson random
variables can be useful approximations for counting the
number of times some event occurs over a very large number of hands. This chapter discusses some continuous random variables that can also yield useful approximations.

6.1 Probability Density Functions

Recall from the beginning of Chapter 4 that a *continuous*
random variable can take any value in the real line (or
interval of the real line or collection of such intervals).
For instance, if X is the exact time, in minutes, before
you win a hand, then in principle X could be any value
in $[0, \infty]$. Precise values such as 8.734301 or even e or
π are conceivable and it may not be possible to list all

the possible outcomes X could take. For such continuous random variables, one cannot summarize their distribution with a probability mass function, and the continuous analog of the pmf is called the *probability density function* or *pdf*. For a continuous random variable X, a pdf $f(y)$ is a non-negative function such that the integral $\int_a^b f(y)\,dy = P(a \leq X \leq b)$ for any real numbers a and b.

As discussed in Section 1.4, the connection between area and probability is worth noting. According to the definition of the pdf, the *area* under the pdf between a and b is equivalent to the probability that X is between a and b. Figure 6.1a shows an example in which the pdf of X is given by $f(y) = 3/20 - 3y^2/2000$ for $0 \leq y \leq 10$ and $f(y) = 0$ otherwise. The integral $\int_{-\infty}^{\infty} f(y)\,dy = 1$, as it must for any pdf since $P(-\infty \leq X \leq \infty) = 1$.

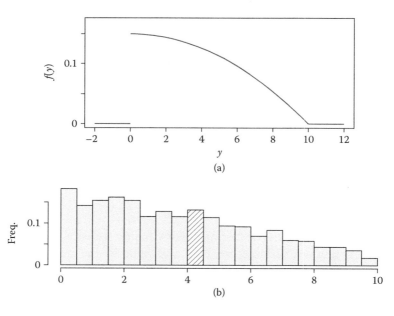

(a)

(b)

FIGURE 6.1 (a) pdf $f(y) = 3/20 - 3y^2/2000$, for $0 \leq y \leq 10$, and $f(y) = 0$ otherwise; and (b) 1000 independent draws from the density $f(y)$.

If a random variable X has pdf f, then the probability for X to assume any particular value c is always zero, since $\int_c^c f(y)dy = 0$ for any c. It is possible in principle for a random variable X to take values in a continuum such as $[0,10]$ and yet have a positive probability of assuming certain discrete values, but such variables arise rarely in practical applications, and in such cases we say the pdf of X does not exist. Thus when X does have a pdf,

$$P(a \leq X \leq b) = P(a < X \leq b) = P(a \leq X < b) = P(a < X < b).$$

Example 6.1.1

Suppose the time in minutes until you win your first hand of Texas Hold'em is modeled as having the pdf in Figure 6.1a, where $f(y) = 3/20 - 3y^2/2000$, for $0 \leq y \leq 10$ and $f(y) = 0$ otherwise. According to this model, what is the probability that it will take between 3 and 5 minutes to win your first hand? What is the probability that it will take more than 8 minutes to win a hand? What is the probability that it will take more than 10 minutes to win a hand?

Answer Letting X = the number of minutes until you win a hand,

$$P(3 \leq X \leq 5) = \int_3^5 (3/20 - 3y^2/2000)dy$$

$$= [3y/20 - y^3/2000]_{y=3}^{5}$$

$$= 15/20 - 125/2000 - 9/20 + 27/2000$$

$$= 25.1\%.$$

$$P(X > 8) = \int_8^\infty f(y)dy$$

$$= \int_8^{10} f(y)dy$$

$$= [3y/20 - y^3/2000]_{y=8}^{10}$$

$$= 30/20 - 1000/2000 - 24/20 + 512/2000$$

$$= 112/2000 = 5.6\%.$$

$$P(X > 10) = \int_{10}^{\infty} f(y)dy = 0.$$

The fact that $P(X > 10) = 0$ according to the quadratic model in Figure 6.1a shows that this model is not ideal for describing the time until you win your first hand, because there is typically a substantial positive probability that it will take you longer than 10 minutes to win a hand. A more appropriate choice of model for waiting times may be the exponential model discussed in Sections 6.2 and 6.4.

Given *iid* observations X_1, X_2, ..., X_n sampled from some unknown distribution, one way to estimate the distribution is via a *relative frequency histogram* such as that shown in Figure 6.1b. The height of each rectangular bin in such a histogram is the proportion of the sample whose values fall in the corresponding range along the x-axis, divided by the width of the bin along the x-axis. For instance, in Figure 6.1b, the relative frequency histogram of an *iid* sample of size $n = 1000$ was drawn from the distribution with pdf $f(y) = 3/20 - 3y^2/2000$, for $0 \le y \le 10$, and 66 of the values, or 6.6%, were between 4 and 4.5. The striped bin stretches along the x-axis from 4 minutes to 4.5 minutes, so the width of this bin is 0.5 minutes, and 6.6% ÷ 0.5 minutes = 0.132 per minute, which is the height of this striped bin along the y-axis. Thus the *area* of the striped rectangular bin is its width times its height = 0.5 minutes × 0.132 per minute = 6.6%, and this indicates the percentage of the sample corresponding to this bin, or the probability that a randomly chosen value from this sample will fall in this bin. With relative frequency histograms, as with probability density functions, *area* corresponds to probability.

6.2 Expected Value, Variance, and Standard Deviation

Recall from Chapter 4 that for a discrete random variable X we define

1. $E(X) = \Sigma \, kP(X = k)$,
2. $\operatorname{var}(X) = E[X - E(X)]^2 = E(X^2) - [E(X)]^2$, and
3. $SD(X) = \sqrt{\operatorname{var}(X)}$.

For a continuous random variable X, the definition of expected value must be changed because $P(X = k) = 0$ for all k. The continuous analog of (1) is $E(X) = \int_{-\infty}^{\infty} y \, f(y) dy$, where f is the pdf of X. Definitions (2) and (3) then apply equally well to discrete or continuous random variables.

Example 6.2.1

Figure 6.2 shows the amounts of profits for players during the first five seasons of *High Stakes Poker*. (The information was taken from the forums of twoplustwo.com and must be incomplete or slightly mistaken since the sum is $700,000, rather than zero.) In Figure 6.2, the curve fitted to the

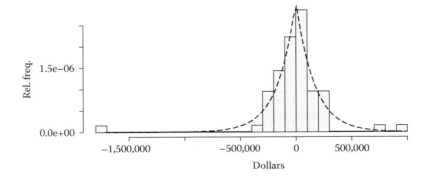

FIGURE 6.2 Profits among players in the first five seasons of *High Stakes Poker*, along with the curve $f(x) = a \exp(-a/x)/2$, with $a = 6.14 \times 10^{-6}$.

data is $f(x) = a \exp(-a|x|)/2$, where $a = 6.14 \times 10^{-6}$. Verify that f is a pdf. If a value X is drawn at random with this pdf f, what are $E(X)$ and $SD(X)$?

Answer For f to be a pdf, it must be non-negative and integrate to one. $f(y)$ is clearly non-negative for all y, and since $f(y) = f(-y)$,

$$\int_{-\infty}^{\infty} f(y)dy = 2\int_0^{\infty} f(y)dy$$
$$= a\int_0^{\infty} \exp(-ay)dy$$
$$= -\exp(-ay)]_0^{\infty}$$
$$= 1.$$

$E(X) = \int_{-\infty}^{\infty} y\, f(y)dy = 0$, since f is symmetric around 0.

$E(X^2) = \int_{-\infty}^{\infty} y^2 f(y)dy$
$$= 2\int_0^{\infty} y^2 f(y)dy$$
$$= a\int_0^{\infty} y^2 \exp(-ay)dy, \text{ which, integrating by parts twice,}$$
$$= [-y^2 \exp(-ay)]_0^{\infty} + 2\int_0^{\infty} y \exp(-ay)dy$$
$$= 0 + 2\int_0^{\infty} y\exp(-ay)dy$$
$$= [-2y \exp(-ay)/a]_0^{\infty} + 2/a\int_0^{\infty} \exp(-ay)dy$$
$$= 0 + 2/a\int_0^{\infty} \exp(-ay)dy$$
$$= [-2/a^2 \exp(-ay)]_0^{\infty}$$
$$= 2/a^2.$$

Thus $\text{var}(X) = E(X^2) - [E(X)]^2 = 2/a^2 = \$53,050,960,753$, so $SD(X) = \sqrt{\text{var}(X)} = \$230,327.90$.

Example 6.2.2

The relative frequency histogram in Figure 6.3 shows the times (including breaks) between eliminations for the final tables at WSOP Main Events

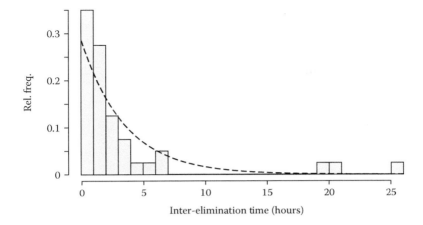

FIGURE 6.3 Times between eliminations in final tables of World Series of Poker Main Events from 2005 through 2010 according to live updates at cardplayer.com, along with the curve $f(x) = a \exp(-ax)$ where $a = 0.285$.

from 2005 to 2010. The information was taken from cardplayer.com's live updates. If X is a randomly selected inter-elimination time in hours, then using the approximation $f(y) = a \exp(-ay)$, with $a = 0.285$, as shown in Figure 6.3, what are $E(X)$ and $\text{var}(X)$?

Answer

$$
\begin{aligned}
E(X) &= \int_{-\infty}^{\infty} y\, f(y)dy \\
&= a\int_0^{\infty} y \exp(-ay)dy \\
&= [-y\exp(-ay)]_0^{\infty} + \int_0^{\infty} \exp(-ay)dy \\
&\quad (\text{employing integration by parts}) \\
&= [0 - \exp(-ay)/a]_0^{\infty} \\
&= 1/a \\
&= 1/0.285 \sim 3.51 \text{ hours.}
\end{aligned}
$$

Integrating by parts again, and using the fact just shown that $a\int_0^{\infty} y \exp(-ay)dy = 1/a$,

$$E(X^2) = \int_{-\infty}^{\infty} y^2 f(y)dy$$
$$= a \int_0^{\infty} y \exp(-ay)dy$$
$$= [-y^2 \exp(-ay)]_0^{\infty} + 2\int_0^{\infty} y \exp(-ay)dy$$
$$= 2/a^2.$$
$$\text{So, } V(X) = E(X^2) - [E(X)]^2$$
$$= 2/a^2 - 1/a^2$$
$$= 1/a^2$$
$$= 1/0.285^2 \sim 12.31 \text{ hours.}$$

The model in Example 6.2.2 is an *exponential distribution*. These distributions, which are often used to describe waiting times, especially for events occurring randomly at a constant rate, are discussed further in Section 6.4. These and some other probability density functions are so commonly used that they deserve special attention, and a selection of these is discussed in the following sections.

6.3 Uniform Random Variables

Perhaps the simplest example of a random variable is the uniform random variable, where the pdf is constant within some specified range $[a,b]$. That is, $f(y) = 1/(b-a)$, for $a \le y \le b$, where a and b are constants with $a \le b$, and $f(y) = 0$ otherwise. Since a uniform random variable X is continuous, $P(X = a) = P(X = b) = 0$, so whether all the inequities above are strict is immaterial.

Notice that $f(y) \ge 0$ for all y, and

$$\int_{-\infty}^{\infty} f(y)dy = \int_a^b f(y)dy = (b-a)/(b-a) = 1,$$

so f is a valid probability density function. If X is a random variable with f as its pdf, we say that X is *uniform on* $[a,b]$, and in this case $E(X) = (a+b)/2$ and $SD(X) = (b-a)/\sqrt{12}$. Indeed,

$$E(X) = \int_a^b y\, f(y)dy$$
$$= (b^2 - a^2)/[2(b - a)]$$
$$= (a + b)/2, \text{ and}$$
$$E(X^2) = \int_a^b y^2 f(y)dy$$
$$= (b^3 - a^3)/[3(b - a)], \text{ so}$$
$$\mathrm{var}(X) = (b^3 - a^3)/[3(b - a)] - (a + b)^2/4$$
$$= (4b^3 - 4a^3 - 3a^2b + 3a^3 - 6ab^2 + 6a^2b - 3b^3 + 3ab^2)$$
$$\div [12(b - a)]$$
$$= (b^3 - a^3 + 3a^2b - 3ab^2)/[12(b - a)]$$
$$= (b - a)^3/[12(b - a)]$$
$$= (b - a)^2/12.$$

If X is a uniform random variable on (a,b), then its moment-generating function is

$$\phi_X(t) = E(e^{tX})$$
$$= \int_{-\infty}^{\infty} e^{ty} f(y)dy$$
$$= \int_a^b e^{ty}[1/(b - a)]dy$$
$$= (e^{tb} - e^{ta})/[t(b - a)].$$

Example 6.3.1

Suppose you play in a large tournament with $n = 100$ other participants. Let X be the *percentage* of your opponents who are eliminated before you are eliminated. If everyone plays equivalently, then X is equally likely to be any of the values 0, 1, 2,..., 100. Only 100 discrete values are possible, but we can see that, as $n \to \infty$, the distribution of X will approach the uniform distribution on [0,100].

For large n, using the uniform distribution as an approximation, what is $P(X \text{ } is \text{ } between \text{ } 20 \text{ } and \text{ } 35)$? What are $E(X)$ and $SD(X)$?

Answer Using the approximation that $X \sim uniform$ on $[0,100]$, the pdf of X is

$f(y) = 1/100$ for $0 \le y \le 100$ and $f(y) = 0$ for all other y, so

$$P(20 \le X \le 35) = \int_{20}^{35} f(y)\,dy = (35 - 20)/100 = 0.15.$$

$$E(X) = (100 + 0)/2 = 50,$$

$$\text{and } SD(X) = (100 - 0)/\sqrt{12} \sim 28.9.$$

Example 6.3.2

Suppose X and Y are independent uniform $(0,1)$ random variables. Let $Z = max\{X,Y\}$. Find (1) the pdf of Z, (2) the expected value of Z, and (3) the standard deviation of Z.

Answer

1. First, consider $F(c)$, the cdf of Z, and note that $Z \le c$ if and only if both $X \le c$ and $Y \le c$. Since X and Y are independent,

$$F(c) = P(Z \le c)$$

$$= P(X \le c \text{ and } Y \le c)$$

$$= P(X \le c)\,P(Y \le c)$$

$$= c^2 \text{ for } 0 \le c \le 1.$$

Since $F(c) = \int f(z)dz$, where $f(z)$ is the pdf of Z, from the fundamental theorem of calculus, we have $f(c) = F'(c) = 2c$, for $0 \le c \le 1$.

2. $E(Z) = \int_0^1 c\, f(c)\, dc = \int_0^1 2c^2 dc = 2/3.$
3. $E(Z^2) = \int_0^1 c^2 f(c)\, dc = \int_0^1 2c^3 dc = 1/2,$

so $V(Z) = E(Z^2) - [E(Z)]^2 = (1/2) - (2/3)^2 = 1/18,$ and $SD(Z) = \sqrt{(1/18)} \sim 0.2357.$

Example 6.3.3

The uniform random variable arises in studying optimal strategy in simplified versions of poker. For instance, in *The Mathematics of Poker*, Chen and Ankenman (2006) analyze various heads-up games with players A and B where each gets, instead of two hole cards, a single number independently drawn from a uniform distribution on [0,1]. Let a and b denote their numbers, respectively. In a simplified scenario discussed on p. 115 of Chen and Ankenman (2006), the pot size is initially zero, and player A must check and must call if player B bets, while player B may either check or bet 1 chip. So, player A has no decision to make, and player B must decide whether to bet. Suppose that after the betting, the player with the higher number wins. Let X denote the number of chips player B profits on the hand. If player B attempts to maximize her expected profit, what is her optimal strategy and what is her expected profit from the game?

Answer If player B sees that she has been dealt the number b and decides to bet, then

$$E(X \mid b) = (-1)\, P(a > b) + (1)P(a < b)$$

$$= (-1)(1 - b) + (1)(b)$$

$$= 2b - 1,$$

since a is uniform on [0,1] and independent of b. If player B decides to check instead, then $E(X \mid b) = 0$. Therefore, player B's expected profit is maximized by betting whenever $2b - 1 > 0$, i.e., when $b > 1/2$, and checking otherwise. With this strategy, player B has a profit of 1 chip when she bets and wins, a profit of -1 chip when she bets and loses, and a profit of 0 when she checks. Due to symmetry and the independence of a and b since a and b are both

$>1/2$, it is equally likely that $a > b$ as that $a < b$. Thus,

$$
\begin{aligned}
E(X) &= (1)\, P(b > 1/2 \text{ and } a < b) + (-1)\, P(b > 1/2 \\
&\quad \text{and } a > b) + (0)\, P(b \le 1/2)
\end{aligned}
$$

$$
\begin{aligned}
&= P(b > 1/2)\, P(a < b \mid b > 1/2) - P(b > 1/2) \\
&\quad P(a > b \mid b > 1/2)
\end{aligned}
$$

$$
= 1/2\, P(a < b \mid b > 1/2) - 1/2\, P(a > b \mid b > 1/2)
$$

$$
\begin{aligned}
&= 1/2\, P(a < b \mid a < 1/2 \text{ and } b > 1/2) \\
&\quad P(a < 1/2 \mid b > 1/2) + 1/2\, P(a < b \mid a > 1/2 \\
&\quad \text{and } b > 1/2)\, P(a > 1/2 \mid b > 1/2) - 1/2 \\
&\quad P(a > b \mid a < 1/2 \text{ and } b > 1/2)\, P(a < 1/2 \\
&\quad \mid b > 1/2) - 1/2\, P(a < b \mid a > 1/2 \text{ and} \\
&\quad b > 1/2)\, P(a > 1/2 \mid b > 1/2)
\end{aligned}
$$

$$
\begin{aligned}
&= 1/2\,(1)\,(1/2) + 1/2\,(1/2)\,(1/2) - 1/2(0)(1/2) \\
&\quad - 1/2(1/2)(1/2)
\end{aligned}
$$

$$
= 1/4.
$$

The scenario in Example 6.3.3 is similar to the simplifications of poker studied by Borel (1938) and von Neumann and Morgenstern (1944). Excellent summaries of these games are given in Ferguson and Ferguson (2003) and Ferguson et al., (2007), who, along with Chen and Ankenman (2006), present analyses of several more realistic extensions of these games as well. In the game Borel (1938) investigated and called *la relance*, each player's hand is represented by a uniform $(0,1)$ random variable as in Example 6.3.3, but in *la relance*, each player must ante 1 chip, and player B cannot check but must either fold or bet some predetermined number of chips. If player B bets, then player A can either call or fold. The model of von Neumann and Morgenstern (1944) is identical to *la relance* except that player B can check rather than fold. This latter version is discussed in the following example.

Example 6.3.4

Consider the simplified game analyzed by von Neumann and Morgenstern (1944). The situation is identical to that in Example 6.3.3, but two chips are already in the pot before player B decides to check or bet. If player B bets a chip, then player A can either call or fold. Suppose again that players A and B attempt to maximize their expected profits. What will the optimal strategies of the two players be?

Answer   First, suppose that player B never bluffs. We will show that this strategy is not optimal for B. Assuming that B does not bluff, B will bet if b is greater than some threshold b^* between 0 and 1, and otherwise B will check. Similarly, if player B bets, then player A will call if a is greater than some threshold a^*, and otherwise A will fold.

```
[  B checks  ] [      B bets        ]
0...................b*...........a*...................1.
[       A folds     ] [   A calls  ]
```

Player A will choose the threshold a^* to optimize her expected profit, by assumption. Suppose B bets. If A folds, her profit is 0. If A calls and wins, which will occur whenever b is between b^* and a^*, her profit is $2 + 1 = 3$. If A calls and loses, which will occur whenever b is between a^* and 1, her profit is -1. Player A will choose the threshold a^* so that her equity from calling is equivalent to her equity from folding, a principle known in game theory as the *indifference principle* (Blackwell and Girshick, 1954). That is, when player B bets, player A will choose a^* so that

$$(3)(a^* - b^*) + (-1)(1 - a^*) = 0,$$
$$\text{i.e., } 3a^* - 3b^* - 1 + a^* = 0,$$
$$\text{i.e., } a^* = [1 + 3b^*]/4.$$

Note that we can write this as $a^* = b^* + (1 - b^*)/4$, and thus, since $b < 1$, this implies that $a^* > b^*$. Note that this strategy without bluffing cannot be optimal for player B, because whenever b is between a^* and b^*, if player A calls, then player B necessarily loses. Therefore, if player B decides to check when b is between a^* and b^* and bet instead when b is between 0 and $(a^* - b^*)$, then player B's equity will necessarily increase, because he will still be betting the same fraction of the time but will win more of the checked hands. Thus, player B must bluff some of the time in order to play optimally.

We may thus infer that player B bluffs when b is between 0 and some number b_1^*, checks when b is between b_1^* and b_2^*, and bets when b is between b_2^* and 1. By the indifference principle, player A will choose the cut-off a^* at which to call by equating the equity of calling with the equity of folding when $a = a^*$. The value b is between 0 and b_1^* with probability b_1^*, in which case A calls and wins three chips, and b is between b_2^* and 1 with probability $1 - b_2^*$, in which case A calls and loses one chip, so the indifference principle leads to

$$0 = (3)(b_1^*) + (-1)(1 - b_2^*), \text{ i.e., } b_2^* = 1 - 3b_1^*.$$

Similarly, player B chooses the thresholds b_1^* and b_2^* so that B is indifferent whether to bet or check when $b = b_1^*$ or b_2^*. If $b = b_1^*$ and player B bets, then he loses one chip if a is between a^* and 1, and he wins three chips otherwise. If $b = b_1^*$ and player B checks, then he loses no chips if $a > b_1^*$ and wins two chips otherwise. Thus,

$$(-1)(1 - a^*) + (2)(a^*) = (0)(1 - b_1^*) + (2)(b_1^*),$$
$$\text{so } a^* = (1 + 2b_1^*)/3.$$

Considering the case $b = b_2^*$, one similarly obtains

$$(-1)(1 - b_2^*) + (3)(b_2^* - a^*) + (2)a^* = (2)b_2^* + (0)$$
$$(1 - b_2^*), \text{ so } a^* = 2b_2^* - 1.$$

Plugging this into the previous equation, we have $2b_2^* - 1 = (1 + 2b_1^*)/3$. Combining this with the previous relation between b_1^* and b_2^* yields $2[1 - 3b_1^*] - 1 = (1 + 2b_1^*)/3$, implying that $b_1^* = 0.1$. This means $b_2^* = 0.7$ and $a^* = 0.4$. Player B bluffs with a probability of 10% and bets for value with the best 30% of hands, while player A calls with the best 60% of hands.

[B bluffs] [B checks] [B bets]

0............b_1^*......a^*.........b_2^*............1.

[A folds] [A calls]

Example 6.3.5

The game analyzed by von Neumann and Morgenstern (1944) is slightly more general than in Example 6.3.4, in that B is allowed to bet any non-negative number c of chips rather than a fixed bet size of 1. If players A and B choose strategies that maximize their expected numbers of chips, then what will these optimal strategies be?

Answer Define the strategic thresholds a^*, b_1^*, and b_2^* as in the solution to Example 6.3.4. We first find a^*, b_1^*, and b_2^* in terms of c. The indifference principle leads to the following equations:

1. $2a^* + (-c)(1 - a^*) = 2b_1^*$.
2. $2a^* + (b_2^* - a^*)(2 + c) + (-c)(1 - b_2^*) = 2b_2^*$.
3. $(2 + c)[b_1^*] + (-c)(1 - b_2^*) = 0$.

Rewriting the first equation yields $a^* = (2b_1^* + c)/(2 + c)$.

Expanding the second equation and solving for b_2^*, we obtain

$$b_2^* = (a^* + 1)/2.$$

The third indifference equation implies $2b_1^* + b_1^*c = c - b_2^*c$.

Removing the asterisks for brevity and solving for b_1, a, and b_2,

$$b_1 = c(1 - b_1)/(2 + c)^2,$$

so

$$b_1 + b_1c/(2 + c)^2 = c/(2 + c)^2.$$

$$b_1(2 + c)^2 + b_1c = c.$$

$$b_1 = c/(c + (2 + c)^2).$$

$$b_1 = c/(c^2 + 5c + 4).$$

$$b_1 = c/((c + 1)(c + 4)).$$

Thus,

$$\begin{aligned}
a &= 1/((2 + c)(c + 1)(c + 4)) \, (c(c + 1)(c + 4) + 2c) \\
&= (c^3 + 5c^2 + 4c + 2c)/((2 + c)(c + 1)(c + 4)) \\
&= c(c^2 + 5c + 6)/((2 + c)(c + 1)(c + 4)) \\
&= c(c + 3)(c + 2)/((c + 1)(c + 2)(c + 4)) \\
&= c(c + 3)/((c + 1)(c + 4)),
\end{aligned}$$

and

$$\begin{aligned}
b_2 &= (a + 1)/2 \\
&= 1/(2(c + 1)(c + 4)) \, ((c + 1)(c + 4) + c(c + 3)) \\
&= (c^2 + 5c + 4 + c^2 + 3c)/(2(c + 1)(c + 4)) \\
&= (2c^2 + 8c + 4)/(2(c + 1)(c + 4)) \\
&= (c^2 + 4c + 2)/(c + 1)(c + 4).
\end{aligned}$$

We now find the optimal value of c that maximizes B's expected profit. Assume the antes in the pot are not counted in terms of player B's profit. That is, if B bets and A folds, then B's profit is 2. (Otherwise, if one subtracts the antes in determining B's profit, then 1 will be subtracted from the profit function below.)

In computing B's expected profit, there are several cases to consider.

If b is in $(0,b_1)$ and a is in $(0,a)$, then the profit for player B is 2.

If b is in $(0,b_1)$ and a is in $(a,1)$, then B's profit = $-c$.

If b is in (b_1,a) and a is in $(0,b_1)$, then B's profit = 2.

If b is in (b_1,a) and a is in $(b1,a)$, then half the time B will win 2 and half the time B will win 0, so the average profit for player B is 1.

If b is in (b_1,a) and a is in $(a,1)$, then B's profit = 0.

If b is in (a,b_2) and a is in $(0,a)$, then B's profit = 2.

If b is in (a,b_2) and a is in (a,b_2), then half the time B will win 2 and half the time B will win 0, so the average profit for player B is 1.

If b is in (a,b_2) and a is in $(b_2,1)$, then B's profit = 0.

If b is in $(b_2,1)$ and a is in $(0,a)$, then B's profit = 2.

If b is in $(b_2,1)$ and a is in (a,b_2), then B's profit = $2 + c$.

If b is in $(b_2,1)$ and a is in $(b_2,1)$, then half the time B will win $2 + c$ and half the time B will win $-c$, so the average profit for player B is 1.

The expected profit for player B is thus

$$b_1a(2) + b_1(1 - a)(-c) + (a - b_1)(b_1)(2) + (a - b_1)^2(1) + 0$$

$$+ (b_2 - a)(a)(2) + (b_2 - a)^2(1) + 0 + (1 - b_2)(a)(2)$$

$$+ (1 - b_2)(b_2 - a)(2 + c) + (1 - b_2)^2(1),$$

which, after some algebra, equals $c/[(c + 1)(c + 4)] + 1$.

Taking the derivative of this profit function $c/[(c + 1)(c + 4)] + 1$ with respect to c and setting it to zero yields $(c + 1)^{-1}(c + 4)^{-1} - c(c + 1)^{-2}(c + 4)^{-1} - c(c + 1)^{-1}(c + 4)^{-2} = 0$, and multiplying by $(c + 1)^2 (c + 4)^2$ yields

$$(c + 1)(c + 4) - c(c + 4) - c(c + 1) = 0,$$
$$c^2 + 5c + 4 - c^2 - 4c - c^2 - c = 0,$$
$$-c^2 + 4 = 0,$$

so $c = 2$. Thus, if player B must use a fixed bet size, then the ideal bet size is the size of the pot.

Plugging $c = 2$ back into the formulas for a, b_1, and b_2, we obtain

$$a = 10/18, \ b_1 = 2/18, \ \text{and} \ b_2{}^* = 14/18.$$

Note that there may be other strategies that are just as good for one player, assuming his opponent plays optimally, but that may be inferior otherwise and are never superior to the solution above. Von Neumann and Morgenstern (1944) call such dominated strategies inadmissible; even if a strategy is optimal against the opponent's optimal strategy, it may be inadmissible if it does not optimally exploit an opponent who chooses a suboptimal strategy.

6.4 Exponential Random Variables

The time until some event occurs is often modeled as an exponential random variable. Recall that in Section 5.3, we discussed geometric random variables for describing the number of iid trials required until some event occurs. The exponential distribution is an extension to the case of continuous time.

Given a fixed parameter $\lambda > 0$ the pdf of an exponential random variable is $f(y) = \lambda \exp(-\lambda y)$, for $y \geq 0$, and $f(y) = 0$ otherwise. Note that $f(y) \geq 0$ for all y and

$$\int_{-\infty}^{\infty} f(y)\,dy = \lambda \int_0^{\infty} \exp(-\lambda y)\,dy = 1,$$

so f is a valid probability density function. If X is exponential with parameter λ, then $E(X) = SD(X) = 1/\lambda$. Indeed, integrating by parts,

$$E(X) = \int_0^{\infty} y\lambda \exp(-\lambda y)\,dy$$

$$= [-y \exp(-\lambda y)]_0^{\infty} + \int_0^{\infty} \exp(-\lambda y)\,dy$$

$$= [-\exp(-\lambda y)/\lambda]_0^{\infty} = 1/\lambda.$$

$$E(X^2) = \int_0^\infty y^2 \lambda \exp(-\lambda y) dy$$

$$= [-y^2 \exp(-\lambda y)]_0^\infty + 2\int_0^\infty y \exp(-\lambda y) dy = 2/\lambda^2.$$

Thus, $\text{var}(X) = 2/\lambda^2 - 1/\lambda^2 = 1/\lambda^2$ and $SD(X) = 1/\lambda$.

The exponential distribution is also closely related to the Poisson distribution discussed in Section 5.5. Recall that if the total numbers of events in any disjoint time spans are independent, then the totals are Poisson random variables. If in addition the events are occurring at a constant rate λ, then the times between events, or *interevent times*, are exponential random variables with mean $1/\lambda$.

Example 6.4.1

Suppose you play 20 hands an hour, with each hand lasting exactly 3 minutes. Let X be the time in hours until the end of the first hand in which you are dealt pocket aces. Use the exponential distribution to approximate $P(X \le 2)$ and compare with the exact solution using the geometric distribution.

Answer Each hand takes 1/20 hours and the probability of being dealt pocket aces on a hand is 1/221 (see Example 2.4.1), so the rate λ at which you expect to be dealt pocket aces is 1 in 221 hands = 1/(221/20) hours ~ 0.0905 per hour. Using the exponential model,

$$P(X \le 2) = 1 - \exp(-2\lambda) \sim 16.556\%.$$

This is an approximation, however, since by assumption X is not continuous and must be an integer multiple of 3 minutes. Let Y = the number of hands you play until you are dealt pocket aces. Using the geometric distribution, $P(X \le 2 \text{ hours})$ = $P(Y \le 40 \text{ hands})$ = $1 - (220/221)^{40} \sim 16.590\%$.

The survivor function for an exponential random variable is particularly simple:

$$P(X > c) = \int_c^\infty f(y)dy$$

$$= \int_c^\infty \lambda \exp(-\lambda y)dy$$

$$= [-\exp(-\lambda y)]_c^\infty$$

$$= \exp(-\lambda c).$$

Like geometric random variables, exponential random variables have the *memorylessness* property discussed in Section 5.3. If X is exponential, then for any non-negative values a and b, $P(X > a + b \mid X > a) = P(X > b)$. To see why, note that

$$P(X > a + b \mid X > a) = P(X > a + b \text{ and } X > a)/P(X > a)$$

$$= P(X > a + b)/P(X > a)$$

$$= \exp[-\lambda(a + b)]/\exp(-\lambda a)$$

$$= \exp(-\lambda b)$$

$$= P(X > b).$$

Thus, with an exponential (or geometric) random variable, if after a certain time you still have not observed the event you are waiting for, then the distribution of the *future*, additional waiting time until you observe the event is the same as the distribution of the *unconditional* time to observe the event to begin with.

Example 6.4.2

Phil Hellmuth, Jr., tends to arrive late and with great fanfare to major tournaments. Suppose you play in a tournament in which 20 hands are dealt per hour, and Hellmuth arrives 5 hours after you start. Using the exponential distribution to approximate the distribution of time it takes to be dealt pocket aces, and assuming you and Hellmuth have no chance of being eliminated before getting pocket aces: (1) what is the probability that you will be dealt pocket aces before Hellmuth arrives? (2) What is the probability that you will be dealt pocket aces before Hellmuth gets pocket aces?

Answer Let X denote the time until you are first dealt pocket aces, and let Y denote the time until Hellmuth gets pocket aces. The probability of being dealt pocket aces on a particular hand is $C(4,2)/C(52,2) = 1/221$, and hands are dealt at a rate of 20 per hour, so the rate λ at which pocket aces are dealt is 20/221 per hour. Approximating X as exponential with rate 20/221,

1. $P(X < 5) = 1 - \exp(-20/221 \times 5)$
   $= 1 - \exp(-100/221) \sim 36.4\%$.

2. $P(X < Y) = P(X < 5) + P(X \geq 5 \text{ and } X < Y)$
   $= P(X < 5) + P(X \geq 5)P(X < Y \mid X \geq 5)$.

$P(X < 5) \sim 36.4\%$ from part (1), so $P(X \geq 5) = 63.6\%$.

Because of the memorylessness of the exponential distribution, given that $X \geq 5$, X and Y have the same distribution, so $P(X < Y \mid X \geq 5) = 1/2$.
Thus $P(X < Y) \sim 36.4\% + 63.6\% (1/2) = 68.2\%$.

6.5 Normal Random Variables

No distribution arises more frequently in applications than the normal distribution, which is often called the *Gaussian distribution* after the German mathematician Carl Friedrich Gauss, who used it to justify his derivation of weighted least-squares estimation. The normal distribution arises naturally as the limiting distribution of *averages* of iid random variables, as discussed in Section 7.2. It has also been used to model numerous other phenomena, especially measurement errors in many different fields.

The probability density function for a normal random variable X is given by

$$f(y) = 1/\sqrt{(2\pi\sigma^2)} \exp\{-(y-\mu)^2/(2\sigma^2)\}$$

for real-valued parameters μ and $\sigma > 0$. The fact that f is a valid probability density function can be verified as follows. First, observe that $f(y) > 0$ for all y. Second, in order to prove that $\int_{-\infty}^{\infty} f(y)dy = 1$, notice that using a change from Cartesian to polar coordinates, i.e., letting $x = r\cos(\theta)$ and $y = r\sin(\theta)$,

$$\int_{-\infty}^{\infty}\int_{-\infty}^{\infty} \exp(-x^2+y^2)/2\,dy\,dx = \int_0^{2\pi}\int_0^{\infty} \exp(-r/2)r\,dr\,d\theta$$
$$= 2\pi\int_0^{\infty} r\exp(-r/2)\,dr$$
$$= -2\pi[\exp(-r/2)]_0^{\infty}$$
$$= 2\pi.$$

Thus, since $\int_{-\infty}^{\infty}\int_{-\infty}^{\infty} \exp\{-(x^2+y^2)/2\}dy\,dx$ can also be written as

$$[\int_{-\infty}^{\infty} \exp(-x^2/2)dx][\int_{-\infty}^{\infty} \exp(-y/2)dy] = [\int_{-\infty}^{\infty} \exp(-y/2)]\,dy]^2,$$

we have $\int_{-\infty}^{\infty} \exp(-y^2/2)dy = \sqrt{(2\pi)}$.
Using the change of variables $z = (x-\mu)/\sigma$,

$$\int_{-\infty}^{\infty} f(y)dy = 1/\sqrt{(2\pi\sigma^2)}\sigma\int_{-\infty}^{\infty} \exp(-z^2/2)\,dz$$
$$= \sigma\sqrt{(2\pi)}/\sqrt{(2\pi\sigma^2)}$$
$$= 1.$$

If X is a normal random variable with parameters μ and σ, then $E(X) = \mu$ and $SD(X) = \sigma$. To see that $E(X) = \mu$, observe that

$$\int_{-\infty}^{\infty} z\exp\{-z^2/(2\sigma^2)\}dz$$
$$= \int_0^{\infty} z\exp\{-z^2/2\sigma^2\}dz + \int_{-\infty}^{-0} z\exp\{-z^2/(2\sigma^2)\}dz$$
$$= \int_0^{\infty} z\exp\{-z^2/(2\sigma^2)\}dz - \int_0^{\infty} z\exp\{-z^2/(2\sigma^2)\}dz$$
$$= 0,$$

so $E(X) = \int_{-\infty}^{\infty} yf(y)dy$

$\quad\quad = \int_{-\infty}^{\infty}(y-\mu)f(y)dy + \int_{-\infty}^{\infty}\mu f(y)dy$

$\quad\quad = 1/\sqrt{(2\pi\sigma^2)}\int_{-\infty}^{\infty} z\exp\{-z^2/(2\sigma^2)\}dz + \mu$

$\quad\quad = \mu,$

where we are using the substitution

$\quad z = y - \mu$

Similarly, we can use the substitution $z = (y - \mu)/\sigma$ to verify that var$(X) = \sigma^2$:

var$(X) = E[(X-\mu)^2]$

$\quad\quad = 1/\sqrt{(2\pi\sigma^2)}\int_{-\infty}^{\infty}(y-\mu)^2\exp\{-(y-\mu)^2/(2\sigma^2)\}dy$

$\quad\quad = \sigma^2/\sqrt{(2\pi)}\int_{-\infty}^{\infty} z^2\exp\{-z^2/2\}dz,$

which, integrating by parts,

$\quad\quad = \sigma^2/\sqrt{(2\pi)}\{-z\exp(-z^2/2)\}]_{-\infty}^{\infty}$

$\quad\quad\quad + \sigma^2/\sqrt{(2\pi)}\int_{-\infty}^{\infty}\exp\{-z^2/2\}dz$

$\quad\quad = \sigma^2/\sqrt{(2\pi)}\{0\} + \sigma^2/\sqrt{(2\pi)}\{\sqrt{(2\pi)}\}$

$\quad\quad = \sigma^2.$

Normal random variables are so commonly used in practice that one often sees the abbreviation $N(\mu, \sigma^2)$ to describe a normal random variable with mean μ and variance σ^2. In the particular case where $X \sim N(0,1)$, we say X is *standard normal*. It is easy to see from the probability density function that if X is normally distributed, then so are aX and $X + a$, for any constant $a \neq 0$. We saw in Chapter 4 that multiplication of a random variable by a constant a results in the expected value multiplied by a and the variance multiplied by a^2, and addition by a results in the expected value increased by a and the variance remaining unchanged. As a result, if $X \sim N(\mu, \sigma^2)$ and $Y = (X - \mu)/\sigma$, then Y is standard normal.

Some features of the standard normal distribution are worth noting. First, the standard normal pdf is symmetric around 0, so for a standard normal random variable Z, $P(Z < 0) = 50\%$, and for any real number a, $P(Z < -a) = P(Z > a)$. In addition, $P(|Z| < 0.674) \sim 50\%$, $P(|Z| < 1) \sim 68.27\%$, and $P(|Z| < 1.96) \sim 95\%$. In other words, any normal random variable is within $0.674\ SD$ of its expected value about 50% of the time, is within $1\ SD$ of its expected value about 68.27% of the time, and is within $1.96\ SD$ of its expected value about 95% of the time.

Example 6.5.1

Suppose that a certain player's profits in a cash game of Texas Hold'em are iid and normally distributed with a SD of \$100 per session, and the player's profit is positive in 75% of the sessions. What is the player's expected profit per session?

Answer Let X denote the player's profit in a given session in dollars. $Z = (X - \mu)/\sigma \sim N(0,1)$.

$$75\% = P(X > 0) = P(Z > -\mu/100),$$

so

$$P(Z \le -\mu/100) = 100\% - 75\% = 25\%,$$

and by the symmetry of the standard normal distribution,

$$P(Z \ge \mu/100) = 25\% \text{ as well.}$$

Hence

$$P(|Z| < \mu/100) = 100\% - 25\% - 25\% = 50\%,$$

as illustrated in Figure 6.4. Recalling that $P(|Z| < 0.674) \sim 50\%$, we see that

$$\mu/100 = 0.674, \text{ so } \mu = 100 \times 0.674 = \$67.40.$$

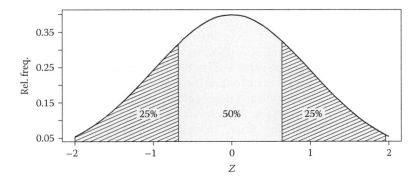

FIGURE 6.4 Areas under standard normal density function.

6.6 Pareto Random Variables

Many phenomena are characterized by the feature that a few of the largest events dwarf all the others in size, and the frequency of these extremely large events is far greater than conventional distributions such as the normal or exponential would predict. Such phenomena are called *heavy-tailed*, because the area under the tail or extreme upper (and/or lower) portion of the distribution is large. Examples include the energy released in earthquakes, the sizes of wildfires, the interevent times or distances between such environmental disturbances, and the population abundances for certain species, including invasive plants or infectious diseases. The Pareto distribution is a heavy-tailed distribution that may be useful for modeling these types of observations. An additional theoretical justification for its use is that the distribution of extreme events such as the maxima of iid observations can under certain conditions be characterized by a Pareto distribution.

The Pareto distribution has cumulative distribution function $F(y) = 1 - (a/y)^b$ and pdf $f(y) = (b/a)(a/y)^{b+1}$ for $y > a$, where $a > 0$ is a *lower truncation point* typically determined or known before observations are recorded, and $b > 0$ is a parameter to be estimated using the data.

Example 6.6.1

After day 5 of the 2010 WSOP Main Event, 205
players remained. Figure 6.5 shows a relative frequency histogram of the top 110 of the chip counts,
along with the pdf of the Pareto distribution with
a = 900,000 and fitted parameter b = 1.11. Suppose
one of these 110 players is selected at random.
Using the fitted Pareto distribution, find the
probability that this player has between 1.5 and
2 million chips left.

Answer   For brevity, let M stand for *million* in
what follows. The probability of the player having less than $2M$ chips is $F(2M) = 1 - (a/2M)^b$ ~
58.78%. The probability of the player having less
than $1.5M$ chips is $F(1.5M) = 1 - (a/1.5M)^b$ ~
43.28%. Thus the probability of the player having
between $1.5M$ and $2M$ chips is $F(2M) - F(1.5M)$ ~
58.78% - 43.28% = 15.50%.

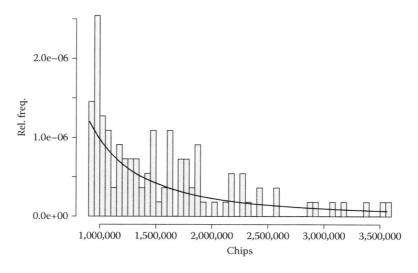

FIGURE 6.5   Relative frequency histogram of the chip counts of the leading
110 players in the 2010 WSOP Main Event after day 5. The curve is the Pareto
density with a = 900,000 and b = 1.11.

The Pareto distribution has a special *self-similarity* or *fractal* property: the shape of the pdf (on log–log scale) looks essentially the same at all scales. To be specific, the probability density function at any value y relative to the density at any other value z depends only on the ratio of y to z and not on the values themselves. The parameter b is often called the *fractal dimension* of the distribution.

Example 6.6.2

Using the Pareto approximation from Example 6.6.1, and again using M to mean *million*, what is the ratio of the probability density of players with $2M$ chips left to the probability density of players with $1M$ chips left? Compare with the ratio of the probability density of players with $4M$ chips left to the probability density of players with $2M$ chips left.

Answer

$$f(2M)/f(1M) = (b/a)(a/2M)^{b+1} \div (b/a)\,(a/1M)^{b+1}$$

$$= (1M/2M)^{b+1} \sim 23.16\%.$$

Similarly, $f(4M)/f(2M) = (2M/4M)^{b+1} \sim 23.16\%$.

A variant of the Pareto distribution is the *tapered* Pareto distribution, which has cumulative distribution function $F(x) = 1-(a/x)^b \exp\{(a-x)/c\}$ and probability density function

$$f(x) = (b/x + 1/c)\,(a/x)^b \exp\{(a-x)/c\} \text{ for } x \geq a,$$

where a, b, $c > 0$. The parameter c is called the *upper cut-off* and governs the exponential taper to 0 of the frequency of large events. Discrimination between the Pareto and tapered Pareto distributions can be difficult in practice because they are so similar, other than in the extreme upper tail, and occurrences in the upper tail are typically rare. The two distributions can have enormously different implications regarding the upper tail, however. For example, both distributions fit similarly to data on energy released in recorded earthquakes worldwide, but while the best-fitting Pareto distribution suggests that an earthquake of magnitude 10.0 or greater should occur every 102 years, the fitted tapered Pareto distribution predicts such an earthquake to happen only every 10^{436} years (Schoenberg and Patel, 2011).

6.7 Continuous Prior and Posterior Distributions

Bayes' rule (see Section 3.4) can be applied to a continuous random variable as a way of updating an estimate of the distribution of an unknown quantity. Suppose observations $Z = \{Z_1, \ldots, Z_n\}$ are to be collected, and that we begin with the knowledge of a *prior* density $f(y)$ for some other quantity X of interest, such that $\int_a^b f(y)dy$ represents the probability of the random variable X falling in the range (a,b) before the data are recorded. Suppose also that, given that X takes the particular value y, the observations $Z = \{Z_1, \ldots, Z_n\}$ have a conditional density $h(Z \mid y)$; this function h is often called the *likelihood function*. Let $g(y|Z)$ represent the *posterior* density of X given Z, i.e., the density such that $\int_a^b g(y|Z)dy$ is the probability of X falling in (a,b), given the recorded data. According to Bayes' rule,

$$g(y|Z) = h(Z \mid y) f(y) \div \int h(Z \mid y) f(y) \, dy.$$

Example 6.7.1

Suppose that, based on your experience, most players in a given card room play about 10% to 30% of the hands they are dealt, and that the frequencies at which different players play hands are approximately uniformly distributed between 10%

and 30%. Assume you have no knowledge about a certain randomly sampled player from the card room. You observe her for 30 hands and find that she plays only 2 of the 30 hands she is dealt. Given this information and assuming that the hands are essentially independent and the player has some constant probability p of playing each hand, what is the posterior probability density for p?

Answer Given that the player's probability of playing a hand is p, the likelihood of Z, the event that she plays exactly 2 of the 30 hands observed, is $C(30,2)\, p^2\, q^{28}$, where $q = 1 - p$. The prior density for p is simply $f(y) = 5$, for y in $(0.1, 0.3)$, and 0 otherwise. Thus, using Bayes' rule, the posterior density for p is

$$g(y|Z) = 5\, C(30,2)\, y^2\, (1-y)^{28}$$

$$\div \int_{.1}^{.3} 5\, C(30,2)\, y^2\, (1-y)^{28}\, dy$$

$$= a\, y^2\, (1-y)^{28} \text{ for } y \text{ in } (0.1, 0.3),$$

where $a = [\int_{.1}^{.3} y^2\, (1-y)^{28}\, dy]^{-1}$, and $g(y|Z) = 0$ otherwise.

An approximation of the posterior density of p is shown in Figure 6.6, along with the prior density of p. One can see that the information that the player has played very few of the last 30 hands results in much higher density corresponding to smaller values of p and very low density near 0.3, compared with the uniform prior density.

Note that it is crucial in the calculation in Example 6.7.1 that the player plays each hand independently with some fixed probability p. In general, this is not the case: if a player has been dealt bad cards and has had to fold many hands in a row, she may feel that her image at the table is that she is especially tight and therefore play the next

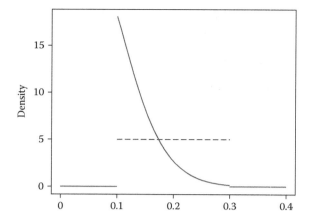

FIGURE 6.6   Prior density (dashed) and posterior density (solid) of the player's handplaying frequency p in Example 6.7.1.

hand with a higher probability. Conversely, some players prefer to play nearly every hand when they first sit down, creating a loose image, and thereafter play only premium hands.

Example 6.7.2

Suppose that, in a given game with no rake, each player i has some long-term average profit μ per hour and that the average profits per hour for different players are normally distributed with mean $0 and *SD* $2. Suppose also that the *SD* of each player's hourly winnings (or losses) is approximately $10. You observe a randomly sampled player from this game over 1 hour and notice that the player wins a total of $50. Given this information, what is the posterior probability density for μ?

Answer   By assumption, the player's hourly profits are normal with mean μ and *SD* $\sigma = \$10$. The likelihood of Z, the event that the player profits $50 in a given hour, is $1/\sqrt{(2\pi\sigma)} \exp\{-(50-\mu)^2/2\sigma^2\} = 1/\sqrt{(20\pi)} \exp\{-(50-\mu)^2/200\}$.

μ is drawn from a normal distribution with mean 0 and *SD* 2, so the prior density of μ is simply $f(y) = 1/\sqrt{(4\pi)}\exp\{-y^2/8\}$. Thus, using Bayes' rule, the posterior density for μ is

$$g(y \mid Z) = 1/\sqrt{(80\pi^2)} \exp\{-y^2/8 - (50-y)^2/200\}$$
$$\div \int_{-\infty}^{\infty} 1/\sqrt{(80\pi^2)} \exp\{-y^2/8 - (50-y)/200\}\, dy$$
$$= a \exp\{-y^2/8 - (50-y)^2/200\},$$

where $a = 1/\int_{-\infty}^{\infty} \exp\{-y^2/8 - (50-y)^2/200\}\, dy$.

Figure 6.7 shows the prior density and posterior density of μ. The information about the player's \$50 win over the last hour shifts the posterior density of μ in the positive direction.

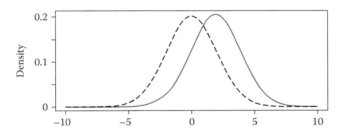

FIGURE 6.7   Prior density (dashed) and posterior density (solid) of long-term hourly profit μ in Example 6.7.2.

Exercises

6.1 Show that the Markov and Chebyshev inequalities proved for discrete random variables in, Section 4.6, hold for continuous random variables as well.

6.2 Suppose each hand lasts exactly 2 minutes, and let X be the time in minutes until the end of the first hand in which you are dealt a pocket pair. Use the exponential distribution to approximate $P(X \le 20)$

and compare with the exact solution using the geometric distribution.

6.3 Show that, for any continuous random variable X with cdf $F(y)$, $F(X)$ is a uniform random variable on $[0,1]$.

6.4 Suppose that your profit per hour in a game of Texas Hold'em is iid and normally distributed with a mean of $10, and that your profit is positive in 75% of the hours you play. What is the standard deviation of your hourly profits?

6.5 Suppose in a year that the poker profits among a group of professional winning players are approximately Pareto distributed. Let X denote the profit of a randomly selected person from this group. If $P(X > \$50,000) = 10\%$ and $P(X > \$25,000) = 80\%$, then what is $P(X > \$100,000)$? What is b?

6.6 Suppose X is Pareto distributed with parameters $a = 1$ and $b = 3$. Find the mean and variance of X.

6.7 Suppose X is Pareto distributed with parameters $a = 1$ and $b > 0$, and let $Y = log(X)$. What is the distribution of Y?

6.8 If different players play with different strategies, then each player may have a different expected proportion of hands that he or she wins. Suppose that each hand is independent of other hands, so that on any hand, player i has probability p_i of winning the hand. Suppose that in a tournament with many participants, the expected proportion of hands won (p_i) for different players is uniformly distributed on $(0.05, 0.15)$. After observing a randomly sampled player (player 1) from this tournament for 70 hands, you find that the player wins 10 of 70 hands. Given only this information, what is the posterior probability density for p_1?

6.9 Suppose X and Y are independent random variables where X is uniform $(0,1)$ and $Y = 1$ with probability $1/3$ and $Y = 2$ with probability $2/3$. If $Z = XY$, find the pdf, expected value, and standard deviation of Z.

6.10 Suppose X and Y are independent uniform $(0,1)$ random variables, and let $Z = min\{X,Y\}$. (a) Find the pdf of Z. (b) Find the expected value of Z.

6.11 Suppose X and Y are independent exponential random variables with parameter λ. What are the pdfs of $max\{X,Y\}$ and $min\{X,Y\}$?

6.12 Suppose X and Y are independent random variables where X is exponential with mean $1/2$, and $Y = 1$ with probability $1/3$ and $Y = 2$ with probability $2/3$. If $Z = XY$, find the pdf, expected value, and standard deviation of Z.

6.13 Show that the moment-generating function of a normal (μ, σ^2) random variable X is $\phi_X(t) = \exp(\mu t + \sigma^2 t^2/2)$. Use this to verify that $E(X) = \mu$ and $var(X) = \sigma^2$.

6.14 Show that the moment-generating function of an exponential (λ) random variable is $\phi_X(t) = \lambda/(\lambda - t)$. Use this to show that $E(X) = 1/\lambda$ and $var(X) = 1/\lambda^2$.

6.15 For the simplified poker game of von Neumann and Morgenstern (1944) analyzed in Example 6.3.5, suppose that three chips (rather than two) are in the pot in the beginning of the hand. (a) Using the indifference principle, write the equations governing the strategy thresholds for players A and B. (b) Find the expected profit per hand for player B, in terms of c and the thresholds in part (a). (c) Show that the profit function from part (b) is maximized when $c = 3$.

6.16 For the von Neumann and Morgenstern (1944) poker game analyzed in Example 6.3.5, suppose that some positive number d of chips (rather than two) are in the pot in the beginning of the hand. (a) Using the indifference principle, write the equations governing the strategy thresholds for players A and B. (b) Find the expected profit per hand for player B, in terms of c, d, and the thresholds in part (a). (c) Show that the profit function from part (b) is maximized when $c = d$. This generalizes Example 6.3.5 to show that, if player B must use a fixed bet size, then player B's ideal bet size is the size of the pot.

6.17 Suppose X is a continuous random variable with cumulative distribution function $F(c) = c^3$, for c between 0 and 1. What is the probability density function of X, for c between 0 and 1?

6.18 Suppose your monthly profits are iid and normally distributed with a *SD* of \$1000, and your profits are positive in 97.5% of the months. What is your expected profit per month?

6.19 Suppose each player in a certain population raises before the flop with some constant frequency, and the frequency is different for different players. Suppose these raising frequencies are normally distributed with mean 0.10 and *SD* 0.025. You observe a randomly sampled player for 100 hands and notice that she raises 15 of the hands. Given this information, what is the posterior probability density for her raising frequency?

CHAPTER 7

Collections of Random Variables

The previous chapters have focused on the description of a single random variable and its associated probabilities. In this chapter, we deal with properties of collections of random variables. Special attention is paid to the probability that the sum or average of a group of random variables falls in some range, and important results in this area include the law of large numbers and the central limit theorem.

7.1 Expected Value and Variance of Sums of Random Variables

Recall from Section 3.1 that two events A and B are independent if $P(AB) = P(A)P(B)$. One similarly refers to *random variables* as independent if the events related to them are independent. Specifically, random variables X and Y are independent if, for any subsets A and B of the real line,

$$P(X \text{ is in } A \text{ and } Y \text{ is in } B) = P(X \text{ is in } A) \, P(Y \text{ is in } B).$$

Technically, the statement above is not exactly correct. For X and Y to be independent, the relation above does not have to hold for all subsets of the real line, but rather for all measurable subsets of the real line. There are some very strange subsets of the real line, such as the Vitali set R/Q, and the probabilities associated with random variables taking values in such sets are not generally defined. These types of issues are addressed in measure-theoretic probability courses.

Even if X and Y are not independent, $E(X + Y) = E(X) + E(Y)$, as long as both $E(X)$ and $E(Y)$ are finite. To see why this is true, suppose X and Y are discrete, and note that

$$P(X = i) = P\{U_j (X = i \text{ and } Y = j)\}$$
$$= \Sigma_j P(X = i \text{ and } Y = j)$$

by the third axiom of probability, and similarly
$P(Y = j) = \Sigma_i P(X = i \text{ and } Y = j)$.
As a result,

$$
\begin{aligned}
E(X + Y) &= \Sigma \, k \, P(X + Y = k) \\
&= \Sigma_i \Sigma_j \, (i + j) \, P(X = i \text{ and } Y = j) \\
&= \Sigma_i \Sigma_j \, iP(X = i \text{ and } Y = j) + \Sigma_i \Sigma_j \, jP(X = i \text{ and } Y = j) \\
&= \Sigma_i \, i \, \{\Sigma_j \, P(X = i, Y = j)\} + \Sigma_j \, j \, \{\Sigma_i \, P(X = i, \, Y = j)\} \\
&= \Sigma_i \, i \, P(X = i) + \Sigma_j \, j \, P(Y = j) \\
&= E(X) + E(Y).
\end{aligned}
$$

A similar proof holds for the case where X or Y or both are not discrete, and it is elementary to see that the statement holds for more than two random variables, i.e., for any random variables $X_1, X_2, \ldots, X_n$, as long as $E(X_i)$ is finite for each i, $E\left(\sum_{i=1}^{n} X_i\right) = \sum_{i=1}^{n} X_i$.

Example 7.1.1

At a 10-handed Texas Hold'em table, what is the expected number of players dealt at least one ace?

Answer   Let $X_i = 1$ if player i has at least one ace, and $X_i = 0$ otherwise.

$E(X_i) = P(\textit{player i is dealt at least one ace})$

$\quad = P(\textit{player i has two aces}) + P(\textit{player i has exactly one ace})$

$\quad = \{C(4,2) + 4 \times 48\}/C(52,2) \sim 14.93\%.$

$\Sigma_i X_i$ = the number of players with at least one ace, and

$$E(\Sigma_i X_i = \Sigma_i E(X_i) = 10 \times 0.1493 = 1.493.$$

Note that, in Example 7.1.1, X_i and X_j are far from independent, for $i \neq j$. If player 1 has an ace, the probability that player 2 has an ace drops dramatically. See Example 2.4.10, where the probability that both players 1 and 2 have at least one ace is approximately 1.74%, so the conditional probability

P(player 2 has at least one ace | player 1 has at least one ace)
  = P(both players have at least one ace) ÷ P(player 1 has at least one ace)

  ~ 1.74%/[1 − C(48,2)/C(52,2)] ~ 11.65%,

whereas the unconditional probability

P(player 2 has at least one ace) = [1 − C(48,2)/C(52,2)]

  ~ 14.93%.

While the sum of expected values is the expected value of the sum, the same is not generally true for variances and standard deviations. However, in the case where the random variables X_i are *independent*,

$$\text{var}\left(\sum_{i=1}^n X_i\right) = \sum_{i=1}^n \text{var}(X_i).$$

Consider the case of two random variables, X and Y.

$$\text{var}(X + Y) = E\{(X + Y)^2\} - \{E(X) + E(Y)\}^2$$

$$= E(X^2) - \{E(X)\}^2 + E(Y^2) - \{E(Y)\}^2 + 2E(XY) - 2E(X)E(Y)$$

$$= \text{var}(X) + \text{var}(Y) + 2[E(XY) - E(X)E(Y)].$$

Now, suppose that X and Y are independent and discrete. Then

$$
\begin{aligned}
E(XY) &= \Sigma_i \, \Sigma_j \, i \, j \, P(X = i \text{ and } Y = j) \\
&= \Sigma_i \, \Sigma_j \, i \, j \, P(X = i) \, P(Y = j) \\
&= \{\Sigma_i \, i \, P(X = i)\} \, \{\Sigma_j \, j \, P(Y = j)\} \\
&= E(X) \, E(Y).
\end{aligned}
$$

A similar proof holds even if X and Y are not discrete, and for more than two random variables: in general, if X_i are *independent* random variables with finite expected value, then $E(X_1 \, X_2 \ldots X_n) = E(X_1) \, E(X_2) \ldots E(X_n)$, and as a result, $\mathrm{var}\left(\Sigma_{i=1}^{n} X_i\right) = \Sigma_{i=1}^{n} \mathrm{var}\left(X_i\right)$ for independent random variables X_i.

The difference $E(XY) - E(X)E(Y)$ is called the *covariance* between X and Y and is labeled $cov(X,Y)$. The quotient $cov(X,Y)/[SD(X) \, SD(Y)]$ is called the *correlation* between X and Y, and when this correlation is 0, the random variables X and Y are called *uncorrelated*.

Example 7.1.2

On one hand during Season 4 of *High Stakes Poker*, Jennifer Harman raised all-in with 10♠ 7♠ after a flop of 10♦ 7♣ K♦. Daniel Negreanu called with K♥ Q♥. The pot was \$156,100. The chances were 71.31% for Harman winning the pot, 28.69% for Negreanu to win and no chance of a tie. The two players decided to *run it twice*, meaning that the dealer would deal the turn and river cards twice (without reshuffling the cards into the deck between the two deals), and each of the two pairs of turn and river cards would be worth half of the pot, or \$78,050. Let X be the amount Harman has after the hand running it twice, and let Y be the amount Harman would have after the hand if they had decided to simply run it once. Compare $E(X)$ to $E(Y)$ and compare approximate values of $SD(X)$ and $SD(Y)$. (In approximating $SD(X)$, ignore the small dependence between the two runs.)

Answer $E(Y) = 71.31\% \times \$156,100 = \$111,314.90$.

If they run it twice, and X_1 = Harman's return from the first run and X_2 = Harman's return from the second run, then

$$X = X_1 + X_2,$$

so

$$E(X) = E(X_1) + E(X_2)$$
$$= \$78,050 \times 71.31\% + \$78,050 \times 71.31\%$$
$$= \$111,314.90.$$

Thus, the expected values of X and Y are equivalent.

For brevity, let B stand for *billion* in what follows:

$$E(Y^2) = 71.31\% \times \$156,100^2 \sim \$17.3B,$$

so

$$V(Y) = E(Y^2) - [E(Y)]^2 \sim \$17.3B - [\$111,314.9^2]$$
$$\sim \$5.09B, \text{ so } SD(Y) \sim \sqrt{\$5.09B} \sim \$71,400.$$

$$V(X_1) = E(X_1^2) - [E(X_1)]^2$$
$$= \$78,050^2 \times 71.31\% - [\$78,050 \times 71.31\%]^2$$
$$\sim \$1.25\ B.$$

Ignoring dependence between the two runs,

$$V(X) \sim V(X_1) + V(X_2) \sim \$1.25B + \$1.25B = \$2.5B,$$

so $SD(X) \sim \sqrt{\$2.5B} = \$50,000$.

Thus, the expected values of X and Y are equivalent ($\$111,314.90$), but the standard deviation of X ($\$50,000$) is smaller than the standard deviation of Y ($\$71,400$).

For independent random variables, $E(XY) = E(X)E(Y)$ (see Exercise 7.10); independent random variables are always uncorrelated. The converse is not always true, as shown in the following example.

Example 7.1.3

Suppose you are dealt two cards from an ordinary deck. Let X = the number on your first card (ace = 14, king = 13, queen = 12, etc.), and let $Y = X$ or $-X$, depending on whether your second card is red or black, respectively. Thus, for instance, $Y = 14$ if and only if your first card is an ace and your second card is red. Are X and Y independent? Are they uncorrelated?

Answer   Consider for instance the events ($Y = 14$) and ($X = 2$).

$P(X = 2) = 1/13$, and counting permutations,

$P(Y = 14) = P$(*first card is a black ace and second card is red*) + P(*first card is a red ace and second card is red*)

$$= (2 \times 26)/(52 \times 51) + (2 \times 25)/(52 \times 51)$$

$$= 102/(52 \times 51)$$

$$= 1/26.$$

X and Y are clearly not independent, since for instance

$$P(X = 2 \text{ and } Y = 14) = 0,$$

whereas

$$P(X = 2)P(Y = 14) = 1/13 \times 1/26.$$

Nevertheless, X and Y are uncorrelated because

$$E(X)E(Y) = 8 \times 0 = 0,$$

and

$$E(XY) = 1/26 \,[(2)(2) + (2)(-2) + (3)(3) + (3)(-3) + \dots + (14)(14) + (14)(-14)]$$

$$= 0.$$

7.2 Conditional Expectation

In Section 3.1, we discussed conditional probabilities in which the conditioning was on an event A. Given a discrete random variable X, one may condition on the event $\{X = j\}$, for each j, and this gives rise to the notion of conditional expectation. A useful example to keep in mind is where you have pocket aces and go all-in, and Y is your profit in the hand, conditional on the number X of players who call you. This problem is worked out in detail, under certain assumptions, in Example 7.2.2. First we will define *conditional expectation*.

If X and Y are discrete random variables, then $E(Y|X=j) = \sum_k k\, P(Y = k\,|\,X = j)$, and the conditional expectation $E(Y\,|\,X)$ is the random variable such that $E(Y\,|\,X) = E(Y\,|\,X = j)$ whenever $X = j$. We will only discuss the discrete case here, but for continuous X and Y the definition is similar, with the sum replaced by an integral and the conditional probability replaced by a conditional pdf. Note that

$$E\{E[Y\,|\,X\,]\} = \sum_j E(Y\,|\,X = j)\, P(X = j)$$

$$= \sum_j \sum_k k\, P(Y = k\,|\,X = j)\, P(X = j)$$

$$= \sum_j \sum_k k\, [P(Y = k \text{ and } X = j)/P(X = j)]\, P(X = j)$$

$$= \sum_j \sum_k k\, P(Y = k \text{ and } X = j)$$

$$= \sum_k \sum_j k\, P(Y = k \text{ and } X = j)$$

$$= \sum_k k \sum_j P(Y = k \text{ and } X = j)$$

$$= \sum_k k\, P(Y = k).$$

Thus, $E\{E[Y\,|\,X]\} = E(Y)$.

Note that using conditional expectation, one could trivially show that the two random variables in Example 7.1.3 are uncorrelated. Because in this example $E[Y\,|\,X]$ is obviously 0 for all X, $E(XY) = E[E(XY\,|\,X)] = E[X\, E(Y\,|\,X)] = E[0] = 0$.

Example 7.2.1

Suppose you are dealt a hand of Texas Hold'em. Let X = the number of red cards in your hand and Y = the number of diamonds in your hand. (1) What is $E(Y)$? (2) What is $E[Y \mid X]$? (3) What is $P\{E[Y \mid X] = 1/2\}$?

Answer

1. $E(Y) = (0)P(Y = 0) + (1)P(Y = 1) + (2)P(Y = 2)$
   $= 0 + 13 \times 39/C(52,2) + 2C(13,2)/C(52,2)$
   $= 1/2.$

2. Obviously, if $X = 0$, then $Y = 0$ also, so

$$E[Y \mid X = 0] = 0,$$

and if $X = 1$, $Y = 0$ or 1 with equal probability, so

$$E[Y \mid X = 1] = 1/2.$$

When $X = 2$, we can use the fact that each of the $C(26,2)$ two-card combinations of red cards is equally likely and count how many have 0, 1, or 2 diamonds. Thus,

$$P(Y = 0 \mid X = 2) = C(13,2)/C(26,2) = 24\%,$$

$$P(Y = 1 \mid X = 2) = 13 \times 13/C(26,2) = 52\%,$$

$$\text{and } P(Y = 2 \mid X = 2) = C(13,2)/C(26,2) = 24\%.$$

So, $E[Y \mid X = 2] = (0)(24\%) + (1)(52\%) + (2)(24\%) = 1$. In summary,

$$E[Y \mid X] = 0 \text{ if } X = 0,$$

$$E[Y \mid X] = 1/2 \text{ if } X = 1,$$

$$\text{and } E[Y \mid X] = 1 \text{ if } X = 2.$$

3. $P\{E[Y \mid X] = 1/2\} = P(X = 1)$

$$= 26 \times 26/C(52,2) \sim 50.98\%.$$

The conditional expectation $E[Y \mid X]$ is actually a random variable, a concept that newcomers can sometimes have trouble understanding. It can help to keep in mind a simple example such as the one above. $E[Y]$ and $E[X]$ are simply real numbers. You do not need to wait to see the cards to know what value they will take. For $E[Y \mid X]$, however, this is not the case, as $E[Y \mid X]$ depends on what cards are dealt and is thus a random variable. Note that, if X is known, then $E[Y \mid X]$ is known too. When X is a discrete random variable that can assume at most some finite number k of distinct values, as in Example 7.2.1, $E[Y \mid X]$ can also assume at most k distinct values.

Example 7.2.2

This example continues Exercise 4.1, which was based on a statement in Volume 1 of *Harrington on Hold'em* that, with AA, "you really want to be one-on-one." Suppose you have AA and go all-in pre-flop for 100 chips, and suppose you will be called by a random number X of opponents, each of whom has at least 100 chips. Suppose also that, given the hands that your opponents may have, your probability of winning the hand is approximately 0.8^X. Let Y be your profit in the hand. What is a general expression for $E(Y \mid X)$? What is $E(Y \mid X)$ when $X = 1$, when $X = 2$, and when $X = 3$? Ignore the blinds and the possibility of ties in your answer.

Answer After the hand, you will profit either $100X$ chips or -100 chips, so $E(Y \mid X) = (100X)$ $(0.8^X) + (-100)(1 - 0.8^X) = [100(X + 1)] \, 0.8^X - 100$. When $X = 1$, $E(Y \mid X) = 60$, when $X = 2$, $E(Y \mid X)$ $= 92$, and when $X = 3$, $E(Y \mid X) = 104.8$.

Notice that the solution to Example 7.2.2 did not require us to know the distribution of X. Incidentally, the approximation $P(winning\ with\ AA) \sim 0.8^X$ is simplistic but may not be a terrible approximation. Using the poker odds calculator at cardplayer.com, consider the case where you have A♠ A♦ against hypothetical players B, C, D, and E who have 10♥ 10♣, 7♠ 7♦, 5♣ 5♦, and A♥ J♥, respectively. Against only player B, your probability of winning the hand is 0.7993, instead of 0.8. Against players B and C, your probability of winning is 0.6493, while the approximation $0.8^2 = 0.64$. Against B, C, and D, your probability of winning is 0.5366, whereas $0.8^3 = 0.512$, and against B, C, D, and E, your probability of winning is 0.4348, while $0.8^4 = 0.4096$.

7.3 Law of Large Numbers and the Fundamental Theorem of Poker

The law of large numbers, which state that the sample mean of *iid* random variables converges to the expected value, are among the oldest and most fundamental cornerstones of probability theory. The theorems date back to Gerolamo Cardano's *Liber de Ludo Aleae* ("Book on Games of Chance") in 1525 and Jacob Bernoulli's *Ars Conjectandi* in 1713, both of which used gambling games involving cards and dice as their primary examples. Bernoulli called the law of large numbers his "Golden Theorem," and his statement, which involved only Bernoulli random variables, has been generalized and strengthened to form the following two law of large numbers. For the following two results, suppose that $X_1, X_2, \ldots$, are *iid* random variables, each with expected value $\mu < \infty$ and variance $\sigma^2 < \infty$.

Theorem 7.3.1 (*weak law of large numbers*)

For any $\varepsilon > 0$, $P(|\overline{X}_n - \mu| > \varepsilon) \to 0$ as $n \to \infty$. ■

Theorem 7.3.2 (*strong law of large numbers*)

$$P(\overline{X}_n \to \mu \text{ as } n \to \infty) = 1.$$

The strong law of large numbers states that for any given $\varepsilon > 0$, there is always some N, so that $|\overline{X}_n - \mu| < \varepsilon$ for all $n > N$. This is a slightly stronger statement and actually implies the weak law, which states that the probability that $|\overline{X}_n - \mu| > \varepsilon$ becomes arbitrarily small but does not expressly prohibit $|\overline{X}_n - \mu|$ from exceeding ε infinitely often. The condition that each X_i has finite variance can be weakened a bit; Durrett (1990), for instance, provides proofs of the results for the case where $E[|X_i|] < \infty$.

Proof.   We will prove only the weak law. Using Chebyshev's inequality and the fact that for independent random variables, the variance of the sum is the sum of the variances,

$$P(|\overline{X}_n - \mu| > \varepsilon) = P\{(\overline{X}_n - \mu)^2 > \varepsilon^2\}$$

$$\leq E\{(\overline{X}_n - \mu)^2\}/\varepsilon^2$$

$$= \text{var}(\overline{X}_n)/\varepsilon^2$$

$$= \text{var}(X_1 + X_2 + \ldots + X_n)/(n^2 \varepsilon^2)$$

$$= \sigma^2/(n \varepsilon^2) \to 0 \text{ as } n \to \infty. \quad ■$$

The law of large numbers are so often misinterpreted that it is important to discuss what they mean and do not mean. The law of large numbers state that for *iid* observations, the sample mean $\overline{X} = (X_1 + \ldots + X_n)/n$ will ultimately converge to the population mean if the sample size gets larger and larger. If one has an unusual run of good or bad luck in the short term, this luck will eventually become negligible in the long run as far as the sample mean is concerned. For instance, suppose you repeatedly play Texas Hold'em and count the number of times you are dealt pocket aces. Let $X_i = 1$ if you get pocket aces on hand i, and $X_i = 0$ otherwise. As discussed in Section 5.1, the sample mean for such Bernoulli random variables is the proportion of hands where you get pocket aces. $E(X_i) = \mu = C(4,2)/C(52,2) = 1/221$, so by the strong law of large numbers, your observed proportion of pocket aces will, with 100% probability, ultimately converge to 1/221. Now, suppose that you get pocket aces on every one of the first 10 hands. This is extremely unlikely, but it has some positive probability ($1/221^{10}$) of occurring. If you continue to play 999,990 more hands, your sample frequency of pocket aces will be

$$(10 + Y)/1 \; million = 10/1{,}000{,}000 + Y/1{,}000{,}000$$

$$= 0.00001 + Y/1{,}000{,}000,$$

where Y is the number of times you get pocket aces in those further 999,990 hands. One can see that the impact on your sample mean of those initial 10 hands becomes negligible.

Note, however, that while the impact of any short-term run of successes or failures has an ultimately negligible effect on the sample mean, the impact on the sample sum does not converge to 0. A common misconception about the law of large numbers is that they imply that for *iid* random variables X_i with expected value μ, the sum

$$(X_1 - \mu) + (X_2 - \mu) + \ldots + (X_n - \mu)$$

will converge to 0 as $n \to \infty$, but this is not the case. If this were true, a short-term run of bad luck, i.e., unusually small values of X_i, would necessarily be counterbalanced later by an equivalent run of good luck, i.e., higher values of X_i than one would otherwise expect and vice versa, but this contradicts the notion of independence. It may be true that due to cheating, karma, or other forces, a short run of bad luck results in a luckier future than one would otherwise expect, but this would mean the observations are not independent and thus certainly has nothing to do with the law of large numbers.

It may at first seem curious that the convergence of the sample mean does not imply convergence of the sample sum. Note that the fact that $\sum_{i=1}^{n}(X_i - \mu)/n$ converges to 0 does not mean that $\sum_{i=1}^{n}(X_i - \mu)$ converges to 0. For a simple counterexample, suppose that $\mu = 0$, $X_1 = 1$, and $X_i = 0$ for all other i. Then the sample sum $\sum_{i=1}^{n}(X_i - \mu) = 1$ for all n, while the sample mean $\sum_{i=1}^{n}(X_i - \mu)/n = 1/n \to 0$. Given a short-term run of bad luck, i.e., if $\sum_{i=1}^{100}(X_i - \mu) = -50$, for instance, while the expected value of the sample mean after 1 million observations is

$$E\left[\sum_{i=1}^{1,000,000}(X_i - \mu)/1,000,000 \mid \sum_{i=1}^{100}(X_i - \mu) = -50\right]$$
$$= -50/1,000,000$$
$$= 0.00005,$$

the expected value of the sum after 1 million observations is

$$E\left[\sum_{i=1}^{1,000,000}(X_i - \mu) \mid \sum_{i=1}^{100}(X_i - \mu) = -50\right]$$
$$= -50 + E\left[\sum_{i=51}^{1,000,000}(X_i - \mu)\right]$$
$$= -50.$$

Again, short-term bad luck is not cancelled out by good luck; it merely becomes negligible when considering the sample mean for large n.

If each observation X_i represents profit from a hand, session, or tournament, then a player is often more interested, at least from a financial point of view, in the sum ΣX_i than the sample mean. The fact that good and bad luck do not necessarily cancel out can thus be difficult for poker players to accept, especially for games like

Texas Hold'em where the impact of luck may be so great. It is certainly true that if $\mu > 0$ and the sample mean $\overline{X}$ converges to μ, then the sample sum ΣX_i obviously diverges to ∞. Thus, having positive expected profits is good because it implies that any short-term bad luck will be dwarfed by enormous long-term future winnings if one plays long enough, but not because bad luck will be cancelled out by good luck.

A related misconception is the notion that, because the sample mean converges to the expected value μ, playing to maximize expected equity is synonymous with obtaining profits. This misconception has been expressed in what David Sklansky, an extremely important author on the mathematics of poker, calls the *fundamental theorem of poker*. Sklansky and Miller (2006) state:

> Every time you play a hand differently from the way you would have played it if you could see all your opponents' cards, they gain; and every time you play your hand the same way you would have played it if you could see all their cards, they lose. Conversely, every time opponents play their hands differently from the way they would have if they could see all your cards, you gain; and every time they play their hands the same way they would have played if they could see all your cards, you lose.(p. 17)

Sklansky (1989) and Sklansky and Miller (2006) provide numerous examples of the fundamental theorem's implications in terms of classifying poker plays as correct or mistakes, where a *mistake* is any play that differs from how the player would (or should) have played if the opponents' cards were exposed.

The basic idea that maximizing your equity may be a useful goal in poker, for the purposes of maximizing future expected profits or for maximizing the probability of winning winner-take-all tournaments was expressed in Section 4.1. However, the fundamental theorem of poker, as framed above, is objectionable for a number of reasons, some of which are outlined below.

1. A theorem is a clear, precise, mathematical statement for which a rigorous proof may be provided. The proof of the fundamental theorem of poker, however, is elusive. The conclusion that you profit every time is not clear, nor are the conditions under which the theorem is purportedly true. Sklansky is most likely referring to the law of large numbers, for which you must assume that your hands of poker form an infinite sequence of *iid* events, and the conclusion is then that your long-term average will ultimately converge to your expected value. It is not obvious that gaining "every time" refers to the long-term convergence of your sample mean.

2. Life is finite, bankrolls are finite, and the variance of no-limit Texas Hold'em is high. If you are only going to play finitely many times in your life, you may lose money in the long term even if you play perfectly. If you have only a finite bankroll, you have a chance to ultimately lose everything even if you play perfectly. This is especially true if you keep escalating the stakes at which you play. One may question whether the law of large numbers really applies to games like tournament Texas Hold'em, where variances may be so large that talking about long-term averages may make little sense for the typical human lifetime.

3. The independence assumption in the law of large numbers may be invalidated in practice. In particular, it may be profitable in some cases to make a play resulting in a large loss of equity if it will give your opponents a misleading image of you and thus give you a big edge later. There is psychological gamesmanship in poker, and in some circumstances you might play a hand "the same way you would have played it if you could see all their cards" and not

gain as much as you would by playing in a totally different, trickier way.

4. A strict interpretation of the conclusion is obviously not true. Sklansky even gives a counterexample right after the statement, "where you have KK on the button with 2.5 big blinds, everyone folds to you, and you go all-in, not realizing that the big blind has AA." Sklansky classifies this as a mistake, although it is obviously the right play. As a group, players in this situation who go all-in will probably make more money than those who do not.

5. The definition of a mistake is arbitrary. There is no reason why conditioning on your opponents' cards is necessarily correct. There are many unknowns in poker, such as the board cards to come. Your opponents' strategies for a hand are also generally unknown and may even be randomized, as with Harrington's watch (see Example 2.1.4). We may instead define a *mistake* as a play contrary to what you would do if you knew what board cards were coming. Better yet, we might define a mistake as a play contrary to what you would do if you knew your opponents' cards, what cards were to come, and how everyone would bet on future rounds. In this case, if you could play mistake-free poker, you really would win money virtually *every time* you sat down. In addition, you have no guarantee that your opponents would necessarily play in such a way as to maximize their equity if they could see your cards.

6. One may also question whether this theorem is indeed fundamental. Since it is generally impossible to know exactly what cards your opponents hold, why should we entertain the notion of people who always know their opponents' cards as ideal players? Instead, why not focus on your overall strategy versus those of your opponents? One might instead define an *ideal player* as someone who uses optimal

strategy for poker (given that, in poker, one does not know one's opponents' cards) and who effectively adjusts this strategy in special cases by reading the opponents' cards, mindsets, mannerisms, and strategies. It might make more sense to say that the closer you are to this ideal, relative to your opponents, the higher your expected value will generally be, and therefore the higher you expect your long-term winnings to be. My friend and professional poker player Keith Wilson noted that, since it implicitly emphasizes putting your opponent on a specific hand rather than a range of hands, "not only is Sklansky's fundamental theorem neither fundamental nor a theorem, it also seems to be just plain bad advice." Harrington and Robertie, in *Harrington on Hold'em*, Volume 1, make a similar point, stating that even the very best players rarely know their opponents' cards exactly and instead simply put their opponent on a range of hands and play accordingly. (Indeed, when I discussed these criticisms with Keith Wilson, he joked that while the fundamental theorem of poker is neither fundamental nor a theorem, the "of poker" part seems right!)

7. The ideal strategy in certain games may be probabilistic, e.g., in some situations it may be ideal to call 60% of the time and raise 40% (see for instance Example 2.1.4). This seems to contradict the notion of the fundamental theorem, which implicitly seems to classify plays as either *correct* or *mistakes* and thus suggests that you should try to do the *correct* play every time.

In summary, the law of large numbers ensure that if your results are like *iid* draws with mean μ, then your sample mean will converge to μ, and maximizing equity may be an excellent idea for maximizing expected profits in the future, but care is required in converting these into statements of certainty about poker results.

7.4 Central Limit Theorem

The law of large numbers dictate that the sample mean of *n iid* observations always approaches the expected value μ as *n* approaches infinity. The next question one might ask is: *how fast?* In other words, given a particular value of *n*, by how much does the sample mean typically deviate from μ? What is the probability that the sample mean differs from μ by more than some specified amount? More generally, what is the probability that the sample mean falls in some range? That is, what is the *distribution* of the sample mean?

The answer to these questions is contained in the central limit theorem, which is one of the most fundamental, useful, and amazing results in probability and statistics. The theorem states that under general conditions, the limiting distribution of the sample mean of *iid* observations is always the *normal* distribution.

To be more specific, let $Y_n = \sum_{i=1}^{n} X_i/n - \mu$ denote the difference between the sample mean and μ, after *n* observations. The central limit theorem states that, for large *n*, Y_n is distributed approximately normally with mean 0 and standard deviation $(\sigma/\sqrt{n})$. It is straightforward to see that Y_n has mean 0 and standard deviation $\sigma/\sqrt{n}$: X_i are *iid* with mean μ and standard deviation σ by assumption, and as discussed in Section 7.1, the expected value of the sum equals the sum of the expected values. Similarly, the variance of the sum equals the sum of the variances for such independent random variables. Thus,

$$E(Y_n) = E\left\{\sum_{i=1}^{n} X_i/n - \mu\right\}$$
$$= \sum_{i=1}^{n} (E\, X_i)/n - \mu$$
$$= n\mu/n - \mu$$
$$= 0,$$

$$\text{and } \operatorname{var}\left(Y_n\right) = \operatorname{var}\left\{\sum\nolimits_{i=1}^{n} X_i / n - \mu\right\}$$

$$= \sum\nolimits_{i=1}^{n} \operatorname{var}\left(X_i\right) / n^2$$

$$= n\sigma^2 / n^2$$

$$= \sigma^2 / n,$$

$$\text{so } SD\left(Y_n\right) = \sigma / \sqrt{n}.$$

Theorem 7.4.1 (*central limit theorem*)

Suppose that X_i are iid random variables with mean μ and standard deviation σ. For any real number c, as $n \to \infty$,

$$P\left\{a \le \left(\bar{X} - \mu\right) \div \left(\sigma / \sqrt{n}\right) \le b\right\} \to 1 / \sqrt{(2\pi)} \int_a^b \exp\left(-y^2 / 2\right) dy.$$

In other words, the distribution of $\bar{X} - \mu \div \left(\sigma / \sqrt{n}\right)$ converges to the standard normal distribution.

Proof. Let ϕ_X denote the moment-generating function of X_i, and let $Z_n = \left(\bar{X} - \mu\right) \div \left(\sigma / \sqrt{n}\right)$. The moment-generating function of Z_n is

$$\phi_{Z_n}(t) = E(\exp\{tZ_n\})$$

$$= E\left[\exp\left\{t\left(\sum\nolimits_{i=1}^{n} X_i / n - \mu\right) \div \left(\sigma / \sqrt{n}\right)\right\}\right]$$

$$= \exp\left(-\mu \sqrt{n} / \sigma\right) E\left[\exp\left\{t \sum\nolimits_{i=1}^{n} X_i / \left(\sigma \sqrt{n}\right)\right\}\right] \qquad (7.4.1)$$

$$= \exp(-\mu\sqrt{n}/\sigma) E[\exp\{tX_1/(\sigma\sqrt{n})\}] E[\exp\{tX_2/(\sigma\sqrt{n})\}]$$
$$\times \dots \times E[\exp\{tX_n/(\sigma\sqrt{n})\}]$$

$$= \exp(-\mu\sqrt{n}/\sigma) \, [\phi_X\{t/(\sigma\sqrt{n})\}]^n, \qquad (7.4.2)$$

where in going from Equation 7.4.1 to Equation 7.4.2, we are using the fact that the values X_i are independent and that functions of independent random variables are always uncorrelated (see Exercise 7.12); thus for any a,

$$E[\exp\{aX_1 + \ldots + aX_n\}] = E[\exp\{aX_1\}] \times \ldots \times E[\exp\{aX_n\}].$$

Therefore, $\log\{\phi_{Zn}(t)\} = n \log [\phi_X\{t/(\sigma\sqrt{n}\}] - \mu\sqrt{n}/\sigma$. Through repeated use of L'Hôpital's rule, and the facts that $\lim_{n\to\infty} \phi_X\{t/(\sigma\sqrt{n}\} = \phi_X(0) = 1$, $\lim_{n\to\infty} \phi'_X\{t/(\sigma\sqrt{n}\} = \phi'_X(0) = \mu$, and $\lim_{n\to\infty} \phi''_X\{t/(\sigma\sqrt{n}\} = \phi''_X(0) = E(X_i^2)$, we obtain

$\lim_{n\to\infty} \log\{\phi_{Zn}(t)\}$

$= \lim_{n\to\infty} n \log [\phi_X\{t/(\sigma\sqrt{n})\}] - \mu\sqrt{n}/\sigma$

$= \lim_{n\to\infty} \{\log [\phi_X\{t/(\sigma\sqrt{n})\}] - \mu/(\sigma\sqrt{n})\}/(1/n)$

$= \lim_{n\to\infty} - n^2\{-\tfrac{1}{2}tn^{-3/2} \phi'_X \{t/(\sigma\sqrt{n})\}/[\sigma\phi_X\{t/(\sigma\sqrt{n})\}] + \tfrac{1}{2}\mu n^{-3/2}/\sigma\}$

$= \lim_{n\to\infty} \{-t/\sigma \ \phi'_X \{t/(\sigma\sqrt{n})\}]/\phi_X\{t/(\sigma\sqrt{n})\} + \mu/\sigma\}/(-2n^{-1/2})$

$= \lim_{n\to\infty} \{\tfrac{1}{2} t^2/\sigma^2 n^{-3/2} \phi''_X \{t/(\sigma\sqrt{n})\}/\phi_X\{t/(\sigma\sqrt{n})\} - \tfrac{1}{2}t^2/\sigma^2 n^{-3/2} [\phi_X\{t/(\sigma\sqrt{n})\}]^{-2} [\phi'_X \{t/(\sigma\sqrt{n})\}]^2\}/n^{3/2}$

$= \lim_{n\to\infty} t^2/(2\sigma^2) [\phi''_X \{t/(\sigma\sqrt{n})\}]/\phi_X\{t/(\sigma\sqrt{n})\} - [\phi_X\{t/(\sigma\sqrt{n})\}]^{-2} [\phi'_X \{t/(\sigma\sqrt{n})\}]^2]$

$= t^2/(2\sigma^2) [E(X_i^2) - \{E(X_i)\}^2]$

$= t^2/(2\sigma^2)[\sigma^2]$

$= t^2/2$

so $\phi_{Zn}(t) \to \exp(t^2/2)$, which is the moment-generating function of the standard normal distribution (see Exercise 6.13). Since convergence of moment-generating functions implies convergence of distributions as discussed in Section 4.7, the proof is complete. ■

The term $\sigma/\sqrt{n}$ is sometimes called the *standard error* for the sample mean. More generally, a *standard error* is the standard deviation of an *estimate* of some parameter. Given n observations, $X_1, \ldots, X_n$, the sample mean

$$\bar{X} = \sum_{i=1}^{n} X_i/n$$

is a natural estimate of the population mean or expected value μ, and the standard deviation of this estimate is $\sigma/\sqrt{n}$.

The amazing feature of the central limit theorem is its generality: regardless of the distribution of the observations X_i, the distribution of the sample mean will approach the *normal* distribution. Even if the random variables X_i are far from normal, the sample mean for large n will nevertheless be approximately normal.

Consider, for example, independent draws X_i from a binomial $(7, 0.1)$ distribution. Each panel in Figure 7.1 shows a relative frequency histogram of 1000 simulated values of $\bar{X}$, for various values of n. For $n = 1$, of course, the distribution is simply binomial $(7, 0.1)$, which is evidently far from the normal distribution. The binomial distribution is discrete, assumes only non-negative values, and is highly asymmetric. Nevertheless, one can see that the distribution of the sample mean converges rapidly to the normal distribution as n increases. For $n = 1000$, the normal pdf with mean 0.7 and standard deviation $\sqrt{(7 \times 0.1 \times 0.9/1000)}$ is overlaid using the dashed curve.

Example 7.4.1

Figure 7.2 is a relative frequency histogram of the number of players still in the hand when the flop was dealt (1 = hand was won before the flop) for each of the 55 hands from *High Stakes Poker* Season 7, Episodes 1 to 4. The sample mean is 2.62 and the sample *SD* is 1.13. Suppose X_i = the number of players on the flop in hand i and suppose that the

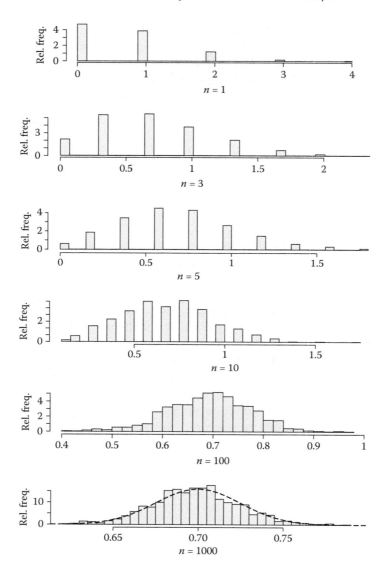

FIGURE 7.1 Relative frequency histograms of the sample mean of *n iid* binomial (7, 0.1) random variables with *n* = 1, 3, 5, 10, 100, and 1000.

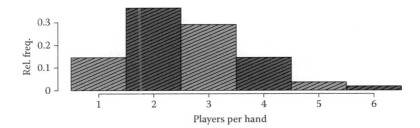

FIGURE 7.2 The number of players per hand on the flop for the 55 hands in the first four episodes of Season 7 of *High Stakes Poker.*

values are *iid* draws from a population with mean 2.62 and *SD* 1.13. Let Y denote the sample mean for the *next* 400 hands. Use the central limit theorem to approximate the probability that $Y \geq 2.6765$.

Answer   According to the central limit theorem, the distribution of Y is approximately normal with mean 2.62 and *SD* $(\sigma/\sqrt{n}) = 1.13/\sqrt{400} = 0.0565$. Thus, $P(Y \geq 2.6765)$ is approximately the probability that a normal random variable takes a value at least 1 *SD* above its mean. To be explicit, if Z is a standard normal random variable, then

$$P(Y \geq 2.6765) = P[(Y - 2.62)/0.0565 \geq (2.6765 - 2.62)/0.0565]$$

$$\sim P[Z \geq (2.6765 - 2.62)/0.0565]$$

$$= P(Z \geq 1).$$

As mentioned in Section 6.5, $P(|Z| < 1) = 68.27\%$, and by the symmetry of the standard normal distribution,

$$P(Z \geq 1) = P(Z \leq -1),$$

so

$$P(Z \geq 1) = \tfrac{1}{2} P(|Z| \geq 1) = \tfrac{1}{2} (100\% - 68.27\%)$$

$$= 15.865\%.$$

Example 7.4.2

Harrington and Robertie (2005) suggest continuation betting on most flops after raising pre-flop and getting only one or two callers because opponents often fold since "most flops miss most hands." Suppose we say the flop *misses* your hand if, after the flop is dealt, you do not have a straight, a flush, an open-ended straight draw, or a flush draw (other than runner–runner draws), nor do any of the numbers on the flop cards match any of the cards in your hand. Over hands referred to in Example 7.4.1, the flop was revealed in 47 of the hands and it missed *all* the players' hands in 11 of the 47 hands. Suppose the probability of the flop missing everyone's hand is always 11/47, and that this event is independent of what happened on previous hands. Let Y denote the proportion of times out of the *next* 100 hands in which the flop is revealed, that the flop misses everyone's hand. Use the central limit theorem to approximate the probability that $Y > 15.10\%$.

Answer Note that the *proportion* of times an event occurs is equal to the *sample mean* if the observations are considered 0s if the event does not occur and 1s if the event occurs, as in the case of Bernoulli random variables. With this convention, each observation has mean $11/47 \sim 23.40\%$ and $SD \sqrt{(11/47)(36/47)} \sim 42.34\%$. Y is the sample mean over the next 100 observations, so by the central limit theorem, Y is distributed like a draw from a normal random variable with mean 23.40% and SD $(42.34\%/\sqrt{100}) = 4.234\%$. Let Z denote a standard normal random variable

$$P(Y > 15.10\%) = P\{(Y - 23.40\%)/4.234\%$$

$$> (15.10\% - 23.40\%)/4.234\%\}$$

$$\sim P(Z > -1.96).$$

As mentioned in Section 6.5, $P(|Z| < 1.96) = 95\%$, and by the symmetry of the standard normal distribution,

$$P(Z \leq -1.96) = P(Z \geq 1.96),$$

so

$$P(Z \leq -1.96) = \tfrac{1}{2} \{P(Z \leq -1.96) + P(Z \geq -1.96)\}$$

$$= \tfrac{1}{2} P(|Z| \geq 1.96)$$

$$= \tfrac{1}{2} (100\% - 95\%) = 2.5\%,$$

and thus $P(Z > -1.96) = 100\% - 2.5\% = 97.5\%$.

7.5 Confidence Intervals for the Sample Mean

In Section 7.4, we saw that the central limit theorem mandates that given n *iid* draws, each with mean μ and SD σ, the difference between the sample mean and μ has a distribution that converges as $n \to \infty$ to normal with mean 0. Examples 7.4.1 and 7.4.2 governed the probability of the sample mean falling in some range, given μ. In this chapter, we will discuss the reverse scenario of finding a range associated with μ based on observation of the sample mean, which is far more realistic and practical. The estimated range that has a 95% probability of containing μ is called a 95% *confidence interval* for μ.

For example, suppose you want to estimate your expected profit μ per hand in a Texas Hold'em game. Assuming the outcomes on different hands are iid, you may observe only a finite sample of n hands, obtain the sample mean over the n hands, and use the central limit theorem to ascertain the likelihood associated with μ falling in a certain range.

Suppose that after $n = 100$ hands, you have profited a total of \$300. Thus, over the 100 hands, your sample mean profit per hand is \$3. Suppose also that the standard deviation of your profit per hand is \$20, and that we may assume the profits on the 100 hands are *iid*. Using the central limit theorem, the quantity $(\overline{X} - \mu)/(\sigma/\sqrt{n})$, which in this case equals $(\$3 - \mu)/(\$20/\sqrt{100}) = \$1.50 - \mu/2$, is standard normally distributed, and thus the probability that this quantity is less than 1.96 in absolute value is approximately 95%. $|1.50 - \mu/2| < 1.96$ if and only if $-0.92 < \mu < 6.92$, i.e., if μ is in the interval $(-0.92, 6.92)$. This interval $(-0.92, 6.92)$ is thus called a *95% confidence interval* for μ. The interpretation is that values in this interval are basically consistent with the results over the 100 hands that were observed.

In general, the 95% *confidence interval* for μ is given by $\left\{\overline{X} - 1.96\,\sigma/\sqrt{n}, \overline{X} + 1.96\,\sigma/\sqrt{n}\right\}$. Confidence intervals can be tricky to interpret. The word *confidence* is used instead of *probability* because μ is not a random variable, so it is not technically correct to say that the probability is 95% that μ falls in some range like $(-0.92, 6.92)$. However, the *interval* $\left\{\overline{X} - 1.96\,\sigma/\sqrt{n}, \overline{X} + 1.96\,\sigma/\sqrt{n}\right\}$ is itself random, since it depends on the sample mean $\overline{X}$, and with 95% probability this random interval will contain μ. In other words, if one were to observe samples of size 100 repeatedly, then for each sample one may construct the confidence interval $\left\{\overline{X} - 1.96\,\sigma/\sqrt{n}, \overline{X} + 1.96\,\sigma/\sqrt{n}\right\}$, and 95% of the confidence intervals will happen to contain the parameter μ.

We have assumed so far that the *SD* σ is known. Recall that σ is the theoretical *SD* of the random variables X_i. In most cases, it is not known and must be estimated using the data. The *SD* of the sample typically converges rapidly to σ, so one may simply replace σ in the formulas above with the sample *SD*, $s = \sqrt{n\left\{\Sigma(X_i - \overline{X})^2/(n-1)\right\}}$. If n is small, however, then the sample *SD* may deviate substantially from σ, and the distribution of $(\overline{X} - \mu)/(s/\sqrt{n})$ has a distribution called the t_{n-1} *distribution*, which is slightly different from the normal distribution but which very closely approximates the normal distribution when n is sufficiently large. As a rule of thumb, some texts propose that when $n > 30$, the two distributions are so similar that the normal approximation may be used even though σ is unknown.

Example 7.5.1

Tom Dwan, whose online screen name is Durrrr, issued the "Durrrr Challenge" in January 2009 offering $1.5 million to anyone (except Phil Galfond) who could beat him in high stakes heads-up no-limit Texas Hold'em or pot-limit Omaha over 50,000 hands, with blinds of at least $200/$400. If Dwan was ahead after 50,000 hands, the opponent would have to pay Dwan $500,000 (profits for either player over the 50,000 hands would be kept as well) Results of the first 39,000 hands against Patrik Antonius were graphed on Coinflip.com's website and can be seen at http://wildfire.stat.ucla. edu/rick/pkrbook/figures/7.5.1.pdf. Do the results prove that Dwan is the better player?

Answer   Over the 39,000 hands played, Dwan profited about $2 million dollars or approximately $51 per hand. Based on the graph, the sample standard deviation over these hands appears to be about $10,000, so, assuming the results on different hands are *iid*, an approximate 95% confidence interval for Dwan's long-term mean profit per hand would be $51 ± 1.96 ($10,000)/√39,000 ~ $51 ± $99 or the interval (−$48, 150). In other words, although the results are favorable for Dwan, the data are insufficient to conclude that Dwan's long-term average profit is positive; it is still highly plausible that his long-term mean profit against Antonius could be zero or negative.

Because the sum of one's profits is of more immediate and practical interest, many poker players and authors graph their *total* profits over some period, although often more can be learned by inspecting a graph of the *mean* profits. Figure 7.3 shows a simulation of *iid* draws from a normal distribution with mean $51 and *SD* $10,000.

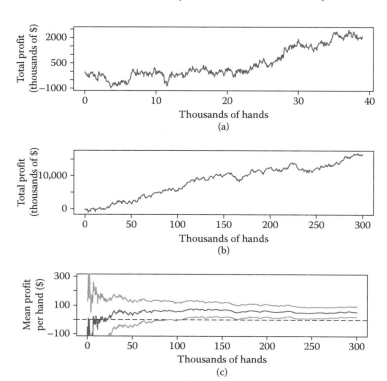

FIGURE 7.3 Sum and average of values drawn independently from a normal distribution with mean $51 and standard deviation $10,000. Panel a shows the sum over the first 40,000 draws. Panel b shows the sum over 300,000 draws. Panel c shows the mean (in red) and 95% confidence intervals for the mean (in pink), over 300,000 draws.

After 40,000 simulated hands, one can see a general upward trend but the graph of the total profits is still characterized by substantial variability. Even after 300,000 simulated hands, the graph of the totals still shows large deviations from a straight line. The graph of the sample mean, however, shows very clear signs of convergence to a positive value after 200,000 hands or so. By the strong law of large numbers, if the simulations were to continue indefinitely, the sample mean would surely converge to 51. Plotting the sample means seems much more instructive than plotting the sample totals.

The 95% confidence interval $\{\bar{X} - 1.96\,\sigma/\sqrt{n},\ \bar{X} + 1.96\,\sigma/\sqrt{n}\}$ is often written simply as $\bar{X} \pm 1.96\,\sigma/\sqrt{n}$, and the quantity $1.96\,\sigma/\sqrt{n}$ is often called the *margin of error*.

Example 7.5.2

Suppose as in the solution to Example 7.5.1 that the *SD* of the profit for Dwan against Antonius on a given hand is $10,000, and that the results on different hands are *iid*. (1) How large a sample must be observed to obtain a margin of error of $10 for the mean profit per hand for Dwan? (2) What about a margin of error of $51?

Answer

1. We want to find n such that $1.96\,(\$10{,}000)/\sqrt{n} = \$10$, i.e., $\sqrt{n} = 1960$, so $n = 3{,}841{,}600$.
2. If $1.96\,(\$10{,}000)/\sqrt{n} = \$51$, then $n = 147{,}697$. Thus, assuming Dwan continues to average about $51 per hand, we would need to observe nearly 148,000 hands (or about 109,000 *more* hands) for our 95% confidence interval for Dwan's true mean profit per hand to be entirely positive.

Note that the sample mean of $51 per hand for Dwan was not needed to find the solution in Example 7.5.2. The margin of error for the sample mean depends on the *SD* and sample size, but not on the sample mean itself.

Example 7.5.3

One statistic commonly recorded by online players is the percentage of hands during which they voluntarily put chips in the pot (VPIP). Most successful players have VPIPs between 15% and 25%. However, over the 70 hands shown on the first six episodes of Season 5 of *High Stakes Poker*, Tom Dwan played 44, so his VPIP was nearly 63% (44/70), yet over these episodes he profited $700,000. (*High Stakes Poker* does not show all hands played, so in reality Dwan's VPIP over all hands may have been considerably lower; ignore this for this example.) Based on these data, find a 95% confidence interval for Dwan's long-term VPIP for this game.

Answer The data can be viewed as Bernoulli random variables with mean 44/70 ~ 62.86%, and such variables have an *SD* of $\sqrt{(62.86\%)(37.14\%)}$ = 48.32%. Thus a 95% confidence interval for the mean percentage of hands in which Dwan's VPIP is 62.86% ± 1.96 (48.32%)/$\sqrt{70}$, or (51.54%, 74.18%).

In Example 7.5.3, the observations X_i are Bernoulli random variables; each is 1 if Dwan enters the pot and 0 if not. For Bernoulli(p) random variables, if p is very close to 0 or 1, the convergence of the sample mean to normality can be very slow. A common rule of thumb for Bernoulli(p) random variables is that the sample mean will be approximately normally distributed if both np and nq are at least 10, where $q = 1 - p$.

7.6 Random Walks and the Probability of Ruin

The previous sections dealt with estimating the probability of a sample mean falling in some range or finding an interval likely to contain μ. We will now cover approximating the distribution of the time before one's chip stack hits zero, the probability that the chip stack stays positive over a certain number of hands, and some related quantities. It is easy to see why such issues would be relevant both to tournament play where the goal is essentially to keep a positive number of chips as long as possible, as well as to cash game play where continued solvency may be as important as long-term average profit. These topics require moving from a description of the random variables $X_0, X_1, \ldots, X_n$, which represent the *changes* to a chip stack or bankroll on hands 0, 1, ..., n, to a description of their *sum* $S_k = \sum_{i=1}^{k} X_i$, for k = 1, 2, ..., n, which form what are called *random walks*. Theoretical results related to random walks, a small sampling of which are provided here, are some of the most fascinating in all of probability theory.

First, a bit of notation and terminology are needed. The treatment here closely follows Feller (1967) and Durrett (2010). Given *iid* random variables X_i, let $S_k = \sum_{i=1}^{k} X_i$, for k = 0, 1, ..., n. This collection of partial sums $\{S_k : k = 0, 1, 2....\}$ is called a *random walk*. The connection of line segments in the plane with vertices (k, S_k), for k = 0 to n is called a *path*. For a *simple* random walk, X_i = 1 or –1, each with probability 1/2, for $i > 0$. Figure 7.4 illustrates a hypothetical example of the path of a simple random walk. In this example, X_0 = 0, but not all simple random walks are required to have this feature.

We should note at the onset that the application of the theory of random walks to actual Texas Hold'em play is something of a stretch. In a typical tournament, for instance, the blinds increase over time, so the outcomes are not *iid*. More importantly, perhaps, the gains and

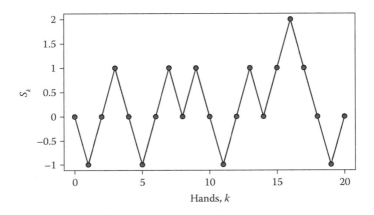

FIGURE 7.4 Sample path of a simple random walk starting at X_0 = 0.

losses on different hands are far from equal and 50–50 propositions, especially for no-limit Texas Hold'em tournaments, in which a player typically endures many small gains and small losses before a large confrontation. Some of the results for random walks may be a bit more relevant to heads-up limit Hold'em. The reader is encouraged to ponder the extension of some of the theory of random walks to more complicated scenarios that may be more applicable to Texas Hold'em, some of which are discussed in Chen and Ankenman (2006).

Note, however, that calculating probabilities involving complex random walks can be tricky. In Chapter 22 of Chen and Ankenman (2006), the authors define the *risk of ruin* function $R(b)$ as the probability of losing one's entire bankroll b. They calculate $R(b)$ for several cases such as where results of each step are normally distributed and for other distributions, but their derivations rely heavily on their assumption that $R(a + b) = R(a)R(b)$, and this result does not hold for their examples, so their resulting formulas are incorrect. If, for instance, the results at each step are normally distributed with mean and variance 1, then using their formula on p. 290, Chen and Ankenman (2006) obtain

$$R(1) = \exp(-2) \sim 13.53\%,$$

but simulations indicate that the probability of the bankroll starting at 1 and reaching 0 or less is approximately 4.15%.

Theorem 7.6.1 (*the reflection principle*)

For a simple random walk, where X_0, n, and y are positive integers, the number of different paths from $(0, X_0)$ to (n, y) that hit the x-axis equals the number of different paths from $(0, -X_0)$ to (n, y).

Proof. The reflection principle may be proven simply by observing the one-to-one correspondence between the two

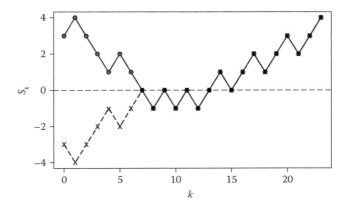

FIGURE 7.5 Reflection principle.

collections of paths; thus their numbers of paths must be equivalent. Figure 7.5 illustrates the one-to-one correspondence. For any path P_1 going from $(0, X_0)$ to (n, y) that hits the x-axis, we may find the *first* time j where the path hits the x-axis and reflect this part of the path across the x-axis to obtain a path P_2 from $(0, -X_0)$ to (n, y). In other words, if P_1 has vertices (k, S_k), then P_2 has vertices $(k, -S_k)$, for $k \leq j$, and for $k > j$, the vertices of P_2 are identical to those of P_1. Thus for each path P_1 from $(0, X_0)$ to (n, y), this correspondence forms a unique path P_2 from $(0, -X_0)$ to (n, y). Similarly, given any path P_2 from $(0, -X_0)$ to (n, y), P_2 must hit the x-axis at some initial time j, and one may thus construct a path P_1 by reflecting the first j vertices of P_2 across the x-axis, i.e., if P_2 has vertices (k, S_k), then P_1 has vertices $(k, -S_k)$, for $k \leq j$, and for $k > j$, the vertices of P_1 and P_2 are identical. ■

Theorem 7.6.2 (*the ballot theorem*)

In $n = a + b$ hands of heads-up Texas Hold'em between players A and B, suppose player A won a of the hands and player B won b hands, where $a > b$. Suppose the hands are then re-played on television in random order. The probability that A has won more hands than B *throughout the telecast* (from the end of the first hand onward) is simply $(a - b)/n$.

Proof. Consider associating each possible permutation of the n hands with the path with vertices (k, S_k), where $S_k = \Sigma_{i=0}^{k} X_i$, $X_0 = 0$, and for $i = 1, \ldots, n$, $X_i = 1$ if player A won the hand and $X_i = -1$ if player B won the hand. Letting $x = a - b$, the number of *different* possible permutations of the n hands is equal to the number of paths from $(0, 0)$ to (n, x), where for two permutations to be considered *different*, the winner of hand i must be different for some i. The number of different permutations is $C(n, a)$, and each is equally likely.

For A to have won more hands than B throughout the telecast, A must obviously win the first hand shown. Thus, letting $x = a - b$, one can see that the number of possible permutations of the n hands where A leads B throughout the telecast is equivalent to the number of paths from $(1, 1)$ to (n, x) that do not touch the x-axis. There are $C(n - 1, a - 1)$ paths from $(1, 1)$ to (n, x) in all. Using the reflection principle, the number of these paths that touch the x-axis equals the number of paths in all going from $(1, -1)$ to (n, x), which is simply $C(n - 1, a)$. Thus, the number of paths from $(1, 1)$ to (n, x) that do not hit the x-axis is

$$C(n - 1, a - 1) - C(n - 1, a) = (n - 1)!/[(a - 1)!(n - a)!] - (n - 1)!/[a!(n - a - 1)!]$$

$$= (n - 1)!/[a!(n - a)!]\{a - (n - a)\}.$$

Thus, the probability of A having won more hands than B throughout the telecast is

$$(n - 1)!/[a!(n - a)!]\{a - (n - a)\} \div C(n, a) = (a - b)/n. \qquad ■$$

Theorem 7.6.2 is often called the *ballot theorem* because it implies that, if two candidates are running for office and candidate A receives a votes and candidate B receives b votes where $a > b$, and if the votes are counted in completely random order, then the probability that candidate A is ahead throughout the counting of the votes is $(a - b)/(a + b)$.

The next theorem shows that, quite incredibly, the probability of avoiding 0 entirely for the first n steps of a simple random walk is equal to the probability of hitting 0 at time n, for any even integer n.

Theorem 7.6.3

Suppose n is a positive even integer, and $\{S_k\}$ is a simple random walk starting at $S_0 = 0$. Then

$$P(S_k \neq 0 \text{ for all } k \text{ in } \{1, 2, \ldots, n\}) = P(S_n = 0).$$

Proof. For positive integers n and j, let $Q_{n,j}$ denote the probability of a simple random walk going from $(0, 0)$ to (n, j). Obviously, $Q_{n,j}$ is also equal to the probability of a simple random walk going from $(1, 1)$ to $(n + 1, j + 1)$, for instance. Now, consider $P(S_1 > 0, \ldots, S_{n-1} > 0, S_n = j)$, for some positive even integer j, which is the probability of going from $(0, 0)$ to $(1, 1)$ and then from $(1, 1)$ to (n, j) without hitting the x-axis. If $j > n$, then this probability is obviously 0, and for $j \leq n$, by the reflection principle, the probability of going from $(1, 1)$ to (n, j) without hitting the x-axis is equal to the probability of going from $(1, 1)$ to (n, j) minus the probability of going from $(1, -1)$ to (n, j), which is simply $Q_{n-1, j-1} - Q_{n-1, j+1}$. Thus, for $j = 2, 4, 6, \ldots, n$,

$$P(S_1 > 0, \ldots, S_{n-1} > 0, S_n = j) = \tfrac{1}{2}(Q_{n-1, j-1} - Q_{n-1, j+1}).$$

By symmetry,

$$P(S_k \neq 0 \text{ for all } k \text{ in } \{1, 2, \ldots, n\})$$

$$= P(S_1 > 0, \ldots, S_n > 0) + P(S_1 < 0, \ldots, S_n < 0)$$

$$= 2P(S_1 > 0, \ldots, S_n > 0)$$

$$= 2\Sigma_{j=2,4,6\ldots,n} \, P(S_1 > 0, \ldots, S_{n-1} > 0, S_n = j)$$

$$= \Sigma_{j=2,4\ldots,n} \, (Q_{n-1, j-1} - Q_{n-1, j+1})$$

$$= [(Q_{n-1, 1} - Q_{n-1, 3}) + (Q_{n-1, 3} - Q_{n-1, 5}) + (Q_{n-1, 5} - Q_{n-1, 7}) + \ldots + (Q_{n-1, n-1} - Q_{n-1, n+1})]$$

$$= Q_{n-1, 1} - Q_{n-1, n+1}$$

$$= Q_{n-1, 1},$$

since $Q_{n-1,n+1} = 0$, because it is impossible for a simple random walk to go from $(0, 0)$ to $(n - 1, n + 1)$. For a standard random walk starting at $S_0 = 0$, $P(S_{n-1} = 1) = P(S_{n-1} = -1)$ by symmetry, and thus

$$P(S_n = 0) = P(S_{n-1} = 1 \text{ and } S_n = 0) + P(S_{n-1} = -1 \text{ and } S_n = 0)$$

$$= P(S_{n-1} = 1 \text{ and } X_n = -1) + P(S_{n-1} = -1 \text{ and } X_n = 1)$$

$$= \tfrac{1}{2} P(S_{n-1} = 1) + \tfrac{1}{2} P(S_{n-1} = -1)$$

$$= P(S_{n-1} = 1)$$

$$= Q_{n-1,1}$$

$$= P(S_k \neq 0 \text{ for all } k \text{ in } \{1, 2, \ldots, n\}). \qquad ■$$

Theorem 7.6.3 implies that, for a simple random walk, for any positive integer n, the probability of avoiding 0 in the first n steps is very easy to compute. Indeed, for such n, $P(S_n = 0)$ is simply $C(n, n/2)/2^n$, which, using Stirling's formula and some calculus, is approximately $1/\sqrt{(\pi n/2)}$ for large n. Thus, if T denotes the positive time when you first hit 0 for a simple random walk, then

$$P(T > n) \sim 1/\sqrt{S(\pi n/2)}.$$

Theorem 7.6.4 (*the arcsine law*)

For a simple random walk beginning at $S_0 = 0$, let L_n be the last time when $S_k = 0$ before time n, i.e., $L_n = max\{k \leq n : S_k = 0\}$. For any interval $[a, b]$ in $[0, 1]$, *as $n \to \infty$*,

$$P(L_{2n}/2n \text{ is in } [a, b]) \to 2/\pi \{arcsin(\sqrt{b}) - arcsin(\sqrt{a})\}.$$

Proof. The reason for the 2s in the expression $L_{2n}/2n$ above and in the expressions below is simply to emphasize that a time when one hits zero must necessarily be an even number. Suppose $0 \leq j \leq n$. Note that $L_{2n} = 2j$ if and only if $S_{2j} = 0$ and then $S_k \neq 0$ for $2j < k \leq 2n$. The key idea in the proof is that one can consider the simple random walk essentially starting over at time $2j$. Thus, by Theorem 7.6.3,

$$P(L_{2n} = 2j) = P(S_{2j} = 0)P(S_{2n-2j} = 0)$$

$$\text{(which, since } P(S_{2j} = 0) \sim 1/\sqrt{(\pi j)})$$

$$\sim 1/\sqrt{(\pi j)}\ 1/\sqrt{[\pi(n-j)]}$$

$$= 1/\{\pi\sqrt{[\ j(n-j)]}\},$$

and thus, if $j/n \to x$, $nP(L_{2n} = 2j) \to 1/\{\pi\sqrt{[x(1-x)]}\}$. Therefore,

$$P(L_{2n}/2n \text{ is in } [a,b]) = \sum_{j:\,a\le j/n\le b} P(L_{2n} = 2j)$$

$$\to \int_a^b 1/\{\pi\sqrt{[x(1-x)]}\}\,dx$$

$$= 2/\pi \int_{\sqrt{a}}^{\sqrt{b}} 1/\sqrt{(1-y^2)}\,dy,$$

employing the change of variables $y = \sqrt{x}$. Using the fact that $\int_{\sqrt{a}}^{\sqrt{b}} 1/\sqrt{(1-y^2)}\,dy = arcsin(\sqrt{b}) - arcsin(\sqrt{a})$, the proof is complete. ▪

Note that, if $a = 0$, $arcsin(\sqrt{a}) = 0$, and if $b = \frac{1}{2}$, $2/\pi$ $arcsin(\sqrt{b}) = \frac{1}{2}$. Thus, for a simple random walk, for large n, the probability that the last time to hit 0 is within the last half of the observations is only 50%. This means that the other 50% of the times, the simple random walk avoids 0 for the last half of the observations. To quote Durrett (2010), "In gambling terms, if two people were to bet \$1 on a coin flip every day of the year, then with probability 1/2 one of the players will be ahead from July 1 to the end of the year, an event that would undoubtedly cause the other player to complain about his bad luck."

As noted above in Theorem 7.6.1, great care must be taken in extrapolating from simple random walks to random walks with *drift*, i.e., where in each step, $E(X_i)$ may not equal 0. One important result related to random walks with drift is the following. In this scenario, at each step i, the gambler must wager some constant fraction a of her total number of chips S_{i-1}. Theorem 7.6.5 is a special case of the more general formula of Kelly (1956), which treats the case where the gambler is receiving *odds* of $b{:}1$ on each wager; i.e., she forfeits aS_{i-1} chips if she loses but gains $b(aS_{i-1})$ if she wins, and her resulting optimal betting ratio a is equal to $(bp - q)/b$. In Theorem 7.6.5, we assume $b = 1$ for simplicity.

Theorem 7.6.5 (*Kelly criterion*)

Suppose $S_0 = c > 0$, and that for $i = 1, 2, \ldots, X_i = aS_{i-1}$ with probability p, and $X_i = -aS_{i-1}$ with probability $q = 1 - p$, where $p > q$ and the player may choose the ratio a of her stack to wager at each step, such that $0 \le a \le 1$. Then as $n \to \infty$, the long-term winning rate $(S_n - S_0)/n$ is maximized when $a = p - q$.

Proof. The following proof is heuristic; for a rigorous treatment, see Breiman (1961). At each step, the player is wagering a times her chip stack, and thus either ends the step with $d(1 + a)$ chips or $d(1 - a)$ chips, if she had d chips before the step. After n steps including k wins and $n - k$ losses, the player will have $c(1 + a)^k (1 - a)^{n-k}$ chips. To find the value of a maximizing this function, set the derivative with respect to a to 0, obtaining

$$0 = ck(1 + a)^{k-1} (1 - a)^{n-k} + c(1 + a)^k (-1)(n - k) (1 - a)^{n-k-1}.$$

$$ck(1 + a)^{k-1} (1 - a)^{n-k} = c(1 + a)^k (n - k) (1 - a)^{n-k-1}.$$

$$k (1 - a) = (1 + a) (n - k).$$

$$k - ak = n - k + an - ak.$$

$$2k = n + an.$$

$$a = (2k - n)/n = 2k/n - 1.$$

Since ultimately $k/n \to p$ by the strong law of large numbers, we find that the long-term optimum is achieved at

$$a = 2p - 1$$

$$= p - (1 - p)$$

$$= p - q.$$ ■

Note that the optimal choice of a does not depend on the starting chip stack c. Theorem 7.6.5 implies that the optimal proportion of stack to wager on each bet is $p - q$. For instance, if one wins in each step with probability $p = 70\%$ and loses with probability 30%, then in order to maximize long-term profit, the optimal betting pattern would be to wager $70\% - 30\% = 40\%$ of one's chips at each step. However, in practice, one typically cannot wager a fraction of a chip, so the application of Theorem 7.6.5 to actual tournament or bankroll situations is questionable, especially when c is small. For further discussion of the Kelly criterion and its applications not only to gambling but also to portfolio management and other areas, see Thorp (1966) and Thorp and Kassouf (1967).

The previous examples are related to the probability of *surviving* for a certain amount of time in a poker tournament. We will end with two results related to the probability of *winning* the heads-up portion of a tournament.

Theorem 7.6.6

Suppose that, heads up in a tournament, there are a total of n chips in play, and you have k of the chips, where k is an integer with $0 < k < n$. You and your opponent will keep playing until one of you has all the chips. For simplicity, suppose also that in each hand, you either gain or lose one chip, each with probability 1/2, and the results of the hands are iid. Then your probability of winning the tournament is k/n.

Proof. The theorem can be proven by induction as follows. Let P_k denote your probability of getting all n chips eventually, given that you currently have k chips. Note first that, trivially, for $k = 0$ or 1, $P_k = kP_1$. Now for the induction step, suppose that for $i = 1, 2, \ldots, j$, $P_i = iP_1$. We will show that, if $j < n$, then $P_{j+1} = (j+1)P_1$, and therefore

$$P_k = kP_1 \text{ for } k = 0, 1, \ldots, n.$$

If you have j chips, for j between 1 and $n - 1$, then there is a probability of 1/2 that after the next hand you will have $j + 1$ chips and a probability of 1/2 that you will have $j - 1$ chips. Either way, your probability of going from there to having all n chips is the same as if you started with $j + 1$ or $j - 1$ chips, respectively. Therefore,

$$P_j = \tfrac{1}{2} P_{j+1} + \tfrac{1}{2} P_{j-1},$$

and by assumption, $P_j = jP_1$ and $P_{j-1} = (j-1)P_1$. Plugging these into the formula above yields

$$jP_1 = \tfrac{1}{2} P_{j+1} + \tfrac{1}{2} (j-1)P_1,$$

and solving for P_{j+1} yields

$$P_{j+1} = 2j\, P_1 - (j-1)P_1$$

$$= (j+1)P_1.$$

By induction, therefore, $P_k = kP_1$ for $k = 0, 1, \ldots, n$. Noting that $P_n = 1$, so $nP_1 = 1$, i.e., $P_1 = 1/n$, the proof is complete. ■

Theorem 7.6.7

Suppose as in Theorem 7.6.6 that, heads up in a tournament, there are a total of m chips in play and you have k of the chips, where now $m = 2^n k$, for some integers k and n, and suppose that in each hand you either double your chips or lose all your chips, each with a probability of 1/2, and the results of the hands are independent. Then as in Theorem 7.6.6, your probability of winning the tournament is equal to your proportion of chips, which is k/m.

Proof. Theorem 7.6.7 can be proven by induction as in the proof of Theorem 7.6.6. Note that for $l = 2^0$, $P_{lk} = lP_k$. Suppose that for $i = 2^0, 2^1, 2^2, \ldots, 2^j$, $P_{ik} = iP_k$.

Thus $P_2^{jk} = 2^j P_k$, and since

$$P_2^{jk} = 1/2\, P_2^{j+1}k + 1/2(0),$$

we have

$$P_2^{j+1}k = 2P_2^{jk} = 2^{j+1}P_k.$$

Therefore, by induction, $P_{ik} = iP_k$, for $i = 2^0, 2^1, \ldots, 2^n$. In particular, $1 = P_{2k}^n = 2^n P_k$, and thus, $P_k = 2^{-n} = k/m$. ■

In the next theorem, instead of each player gaining or losing chips with probability 1/2, the game is skewed slightly in favor of one player.

Theorem 7.6.8

Suppose as in Theorem 7.6.6 that you are heads up in a tournament with a total of n chips in play and you have k of the chips for some integer k with $0 < k < n$. You and your opponent will keep playing until one of you has all the chips. Suppose that in each hand, you either gain one chip from your opponent with probability p or your opponent gains one chip from you with probability $q = 1 - p$, where $0 < p < 1$ and $p \neq 1/2$. Also suppose that the results of the hands are iid. Then your probability of winning the tournament is $(1 - r^k)/(1 - r^n)$, where $r = q/p$.

Proof. The proof here follows Ross (2009). As in the proofs of the two preceding results, let P_k denote your probability of getting all n chips eventually given that you currently have k chips. Starting with k chips, if $1 \leq k \leq n - 1$, after the first hand you will either have $k + 1$ chips (with probability p) or $k - 1$ chips (with probability q). Thus, for $1 \leq k \leq n - 1$,

$$P_k = p\, P_{k+1} + q\, P_{k-1}.$$

Since $p + q = 1$, we may rewrite this as

$$(p + q) P_k = p P_{k+1} + q P_{k-1},$$

or

$$p P_{k+1} - p P_k = q P_k - q P_{k-1},$$

i.e.,

$p(P_{k+1} - P_k) = q(P_k - P_{k-1})$, and letting $r = q/p$, we have

$$P_{k+1} - P_k = r(P_k - P_{k-1}) \text{ for } 1 \le k \le n - 1. \tag{7.6.1}$$

Obviously, $P_0 = 0$, so for $k = 1$, Equation 7.6.1 implies that $P_2 - P_1 = rP_1$. For $k = 2$, Equation 7.6.1 yields

$$P_3 - P_2 = r(P_2 - P_1)$$
$$= r^2 P_1,$$

and one can see that

$$P_{j+1} - P_j = r^j P_1 \text{ for } j = 1, 2, \ldots, n - 1. \tag{7.6.2}$$

Summing both sides of Equation 7.6.2 from $j = 1, 2, \ldots, k - 1$, we obtain

$$P_k - P_1 = P_1 \left(r + r^2 + \ldots + r^{k-1} \right),$$

so

$$P_k = P_1(1 + r + r_2 + \ldots + r_{k-1}),$$

and thus

$$P_k = P_1(1 - r^k)/(1 - r) \text{ for } k = 1, 2, \ldots, n. \tag{7.6.3}$$

For $k = n$, Equation 7.6.3 and the fact that $P_n = 1$ yield

$$1 = P_1 (1 - r^n)/(1 - r),$$

so

$$P_1 = (1 - r)/(1 - r^n). \tag{7.6.4}$$

Combining Equation 7.6.3 with Equation 7.6.4, we have

$$P_k = (1 - r^k)/(1 - r^n), \text{ as desired.} \qquad ■$$

Exercises

7.1 At a 10-handed Texas Hold'em table, what is the expected number of players who are dealt at least one spade?

7.2 Suppose you are dealt a two-card hand of Texas Hold'em. Let X = the number of face cards in your hand and Y = the number of kings in your hand. (a) What is $E(Y)$? (b) What is $E[Y \mid X]$? (c) What is $P\{E[Y \mid X] = 2/3\}$?

7.3 Confrontations like AK against QQ are sometimes referred to as *coin flips* among poker players, even though the player with QQ has about (depending on the suits) a 56% chance of winning the hand. Suppose for simplicity that a winner-take-all tournament with 256 players and an entry fee of $1000 per player is based completely on doubling one's chips and that player X has a 56% chance of doubling up each time because of X's skillful play. (a) What is the probability of X winning the tournament? (b) What is X's expected profit on her investment in the tournament?

7.4 In *Harrington on Hold'em*, Volume 1, Harrington and Robertie (2004) discuss a quantity called M attributed to Paul Magriel. For a 10-handed table, M is defined as a player's total number of chips divided by the total quantity of the blinds plus antes on each hand. (For a short-handed table with k players, M is computed by multiplying the quotient above by $k/10$). Suppose you are in the big blind in a 10-handed cash game with no antes, blinds fixed at $10 and $20 per player per hand, and you have $200 in chips. Suppose you always play each hand with probability $1/M$ and fold otherwise. Find the probability that you lose at least half your chips without ever playing a hand. (Note that M depends on your chip count, which changes after each hand.)

7.5 For a simple random walk starting at 0, let T = the first positive time to hit 0. Compute $P(T > n)$ for $n = 2$ and for $n = 4$, and compare with the approximation $P(T > n) \sim 1/\sqrt{(\pi n/2)}$ from Section 7.6.

7.6 A well-known quote attributed to Rick Bennet is, "In the long run there's no luck in poker, but the short run is longer than most people know." Comment.

7.7 Think of an example of real values (not random variables) x_1, x_2, ..., such that $\lim_n \to \infty \sum_{i=1}^{n} x_i/n = 0$, while $\lim_{n \to \infty} \sum_{i=1}^{n} x_i = \infty$. In one or two sentences, summarize what this example and the law of large numbers mean in terms of the sum and mean of a large number of independent random variables.

7.8 Daniel Negreanu lost approximately $1.7 million in total over the first five seasons of *High Stakes Poker*. However, is he a losing player in this game, or is it plausible that he has just been unlucky and if he were to keep playing this game for a long time, could he be a long-term winner? (a) Find a 95% confidence interval for Negreanu's mean winnings per hand, assuming that his results on different hands are *iid* random variables, that he played about 250 hands per season, and that the *SD* of his winnings (or losses) per hand was approximately $30,000. (b) If Negreanu were to keep losing at the same rate, how many more hands would we have to observe before the 95% confidence interval for Negreanu's mean winnings per hand would be entirely negative, i.e., would not contain 0?

7.9 Suppose X and Y are two independent discrete random variables. Show that $E[Y \mid X]$ is a constant.

7.10 Show that if X and Y are any two independent discrete random variables, then they are uncorrelated, i.e., $E(XY) = E(X)E(Y)$.

7.11 Show that $E(X + Y) = E(X) + E(Y)$ for continuous random variables X and Y with probability density functions f and g, respectively, provided that $E(X)$ and $E(Y)$ are finite.

7.12 Show that, if X and Y are any two independent discrete random variables and f and g are any functions,

$$E[f(X)\,g(Y)] = E[f(X)]\,E[g(Y)].$$

7.13 Suppose as in Theorem 7.6.8 that you are heads up in a tournament with k of the n chips in play where $0 < k < n$, the results on different hands are *iid*, and on each hand you either gain one chip from your opponent with probability p or your opponent gains one chip from you with probability $q = 1 - p$, with $0 < p < 1$ *and* $p \neq 1/2$. In principle, the tournament could last forever, with your chip stack and that of your opponent bouncing between 1, 2, ..., $n - 1$ indefinitely. Find the probability that this occurs, i.e., the probability that neither you nor your opponent ever wins all the chips.

7.14 Suppose you are heads up in a tournament, and that you have two chips left and your opponent has four chips left. Suppose also that the results on different hands are *iid*, and that on each hand with probability p you gain one chip from your opponent, and with probability q your opponent gains one chip from you.

a. If $p = 0.52$, find the probability that you will win the tournament.

b. What would p need to be so that the probability that you will win the tournament is $1/2$?

c. If $p = 0.75$ and your opponent has 10 chips left instead of 4, what is the probability that you will win the tournament? What if your opponent has 1000 chips left?

7.15 Suppose you repeatedly play in tournaments, each with 512 total participants, and suppose that in each stage of each tournament, you either double your chips with probability p or lose all your chips with probability $1 - p$, and the results of the tournament stages are independent. If your probability of winning a tournament is 5%, then what is p?

7.16 Continuing Exercise 3.24, which discussed a hand from day 5 of the 2014 WSOP Main Event where there were six players remaining in the hand and five of them had pocket pairs (Brian Roberts had J♥ J♣, Greg Himmelbrand had 6♠ 6♥, Robert Park had 4♥ 4♣, Adam Lamphere had 10♠ 10♥, and Jack Schanbacher had 10♦ 10♣), let X be 1 if you have a pocket pair and 0 otherwise. Let Y be 1 if the player on your right has a pocket pair and 0 otherwise. (a) What is the covariance between X and Y? (b) What is the correlation between X and Y? (c) What is the expected number of pocket pairs out of six hands? (d) What is the probability that, with six players remaining in the hand, at least five would have pocket pairs? In your answer to part (d), because the correlation you found in part (b) is so small, calculate the probability pretending that events like X and Y are independent.

7.17 Suppose each minute you gain a chip with probability 1/2 or lose one chip with probability 1/2, according to a simple random walk starting with one chip at time 0. Find P(your chip count stays positive for 30 minutes).

7.18 Suppose you have 1000 chips at time 0, and each minute you either gain a chip with probability 1/2 or lose a chip with probability 1/2, according to a simple random walk. You play for 500 minutes and record the times when your chip total is back at 1000. What is the probability that the last time you hit 1000 is between time 100 and 200 minutes?

7.19 The *martingale strategy* refers to a game where, at minute i, for $i = 0, 1, 2, 3, \ldots$, you make a wager of $\$2^i$. Each minute, independently of the others, with probability 1/2 you win (profit) $\$2^i$ and with probability 1/2 you lose your wager of $\$2^i$. You stop when you win for the first time. Let X denote your profit from playing this game once. (a) If you also must stop if you have not won by minute n, find the pdf of X and the expected value of X. (b) As $n \to \infty$, show that the pdf of X converges to $P(X = \$1) = 1$. Does this mean playing the martingale strategy is a good idea?

7.20 Suppose you have \$100 and, each minute, you may wager a certain amount and you will win with probability $p = 0.51$ and lose with probability 0.49, each time independently of what happened in other minutes. Suppose you use the Kelly criterion to determine your wager sizes. How many dollars do you expect to have after 100 minutes?

7.21 Suppose you have 10 chips at time 0, and each minute you either gain a chip with probability 1/2 or lose a chip with probability 1/2, according to a simple random walk. What is the probability that after 40 minutes you have 14 chips and have not hit 0?

## CHAPTER 8

Simulation and Approximation Using Computers

Many probability problems are incredibly cumbersome to solve analytically but can be approximated readily using computer simulations. Imagine, for instance, trying to compute the expected profit of one strategy versus another, the probability of a particular hand like A♠ K♥ winning a showdown against a full table of opponents with random hands, or the probability of a player making a full house or better and losing the hand, given that the players are all-in before they even see their cards.

Rather than spend days or weeks calculating such probabilities with a paper and pencil (and very likely making mistakes in the process), it is possible to approximate the solutions to such problems easily and rapidly using computer simulations. A computer can mimic the shuffling and dealing of cards repeatedly and record each result. After thousands of repetitions, one can typically obtain a sufficiently accurate estimate of the true probability.

Another value of computers is the ability to approximate quantities such as the probability that a draw from a standard normal distribution is less than some value. One can see from Sections 6.4 and 6.5 why such information would be useful. Traditional probability and statistics texts contained tables of normal percentiles and numerous exercises requiring students to compute probabilities or confidence intervals by exacting the appropriate values from these rather complicated tables. Today, the availability of computers to students makes such tables unnecessary.

The statistical programming language called R is free, publicly available, and very easy to use for these purposes. To download and install R, see http://www.r-project.org. Using R, one can find for instance that the probability of a standard normal random variable taking a value less than 1.53 is approximately 93.69916%, simply using the command *pnorm*(1.53).

Example 8.1

In Example 7.4.1, the probability that $Y \geq 2.6765$ was approximated, where Y is the mean number of players still in the hand when the flop was dealt, over the next 400 hands, under the assumption that the numbers of players per hand are iid with mean 2.64 and SD 1.16. Using the central limit theorem, this conveniently worked out to

$$P(Z \geq 1),$$

where Z is a standard normal random variable. Using R, approximate $P(Y \geq 2.8)$.

Answer

$$P(Y \geq 2.8) = P\{(Y - 2.64)/0.058 \geq (2.8 - 2.64)/0.058\}$$

$$\sim P\{Z \geq (2.8 - 2.64)/0.058\}$$

$$\sim P(Z \geq 2.759)$$

$$= 1 - P(Z < 2.759)$$

$$\sim 1 - 0.9971 = 0.29\%,$$

using the R command *pnorm*(2.759).

R can be used readily to mimic drawing a uniform random variable. Such draws are called *pseudo-random*, as they are determined by a computer and are thus not truly random, but are so devoid of visible patterns or violations of uniformity that for many purposes they suffice as approximations to uniformly distributed random variables. In R, one can obtain such a uniform pseudo-random variable on $(0, 1)$ using the command *runif*(1), and, for instance, to obtain 100 variables one can enter *runif*(100).

R is intuitive and many functions can be explored independently or using the manuals available from r-project.org. R also includes easy-to-manipulate and state-of-the-art graphics. The example below illustrates basic R functions, commands, and syntax.

Example 8.2

Simulate 500 randomly scattered points in the diamond with vertices at $(-1, 0)$, $(0, 1)$, $(1, 0)$, and $(0, -1)$.

Answer The diamond specified can be characterized as the set of points $\{(x, y):|x| + |y| < 1\}$. There are multiple ways to simulate 500 points in this diamond, and below is R code that achieves this using a loop. The code proceeds by generating points in the square with vertices $(-1, -1)$, $(1, -1)$, $(1, 1)$, and $(-1, 1)$, and then keeping the first 500 points that fall in the diamond described above.

We first generate a matrix z with 2 columns and 2000 rows, containing all 0s, so that z will be big enough to hold the coordinates of 2000 points. The variable j will represent the number of points simulated in the diamond, which is initially 0. The third line indicates how a loop is initiated in R, and 1:2000 means the list of integers from 1 to 2000. The loop works by generating up to 2000 x-coordinates and y-coordinates randomly between -1 and 1 [since if x and y are uniform $(0, 1)$ random variables, then $2x - 1$ and $2y - 1$ are uniformly distributed on $(-1, 1)$].

The code then proceeds in the sixth line by evaluating whether the point simulated is in the diamond $\{(x, y): |x| + |y| < 1\}$, in which case the coordinates are stored in z. Everything in a line after a # sign is simply a comment.

```
z = matrix(0,ncol=2,nrow=2000)
j=0
for(i in 1:2000){
  x = runif(1)*2-1
  y = runif(1)*2-1
  if(abs(x)+abs(y) < 1){
    z[j,] = c(x,y)
    j = j+1
    }
  }
j # check that j > 500,
z2 = z[1:500,]
plot(c(-1,0,1,0,-1),c(0,1,0,-1,0),
  type ="n",xlab="",ylab ="") # sets
  axes
points(z2,pch=".")
```

The resulting plot is shown in Figure 8.1.

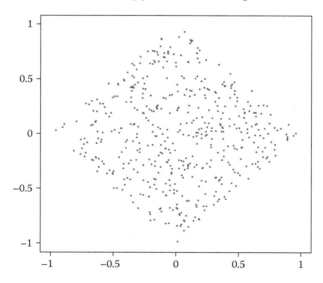

FIGURE 8.1 Five hundred points simulated in the diamond $|x| + |y| < 1$.

R can also be used to simulate poker hands and approximate probabilities related to Texas Hold'em, using the R package called *holdem*. To download and install the package on your computer, use the following R command:

```
install.packages("holdem").
```

To load the functions from this package into your current R session, use the following R command:

```
library(holdem).
```

The *holdem* package contains R functions that may be used to simulate poker hands, as well as examples of Texas Hold'em functions written by former students. For instance, the function *gravity* in *holdem*, which is printed below, was the winning student program from summer 2009; all the students' codes had to go all-in or fold.

The function *gravity* goes all-in with any pair of 10s or higher, or with AK or AQ or AJ, or suited connectors where the lowest card is a 10 or higher. In addition, if the player's total number of chips is smaller than two times the big blind, then *gravity* goes all-in. Otherwise, *gravity* folds. The last argument in a function is what any R function outputs; for the *gravity* function, the output is $a1$, which is its bet size: 0 if *gravity* folds, or its total number of chips if *gravity* goes all-in.

```
gravity = function(numattable1, crds1,
   board1,
   round1, currentbet, mychips1, pot1,
   roundbets, blinds1, chips1, ind1,
     dealer1,
   tablesleft){
## all in with any pair of 10s or
   greater,
## or AJ-AK, or suited connectors with
   lowest
## card a 10 or higher.
```

```
## if your chip count is less than twice
   the
## big blind, go all in
## with any cards.
a1 = 0
if((crds1[1,1] == crds1[2,1]) & &
   (crds1[1,1]
   > 9.5)) a1 = mychips1
if((crds1[1,1] > 13.5) & &
   (crds1[2,1]>10.5))
   a1 = mychips1
if((crds1[1,1]-crds1[2,1]==1) & &
   (crds1[1,2]
   == crds1[2,2]) & & crds1[2,1]>9.5))
      a1 = mychips1
if(mychips1 < 2*blinds1) a1 = mychips1
a1
} ## end of gravity
```

The input variables for the Texas Hold'em strategy functions in *holdem*, such as *gravity*, are the numbers of players at your table (an integer), the cards you have (2 × 2 matrix where the first column represents the numbers of your cards: 2, 3, 4, ..., 10, J = 11, Q = 12, K = 13, A = 14, in order so that the first row's number is always at least as large as the second row's number; the entries in the second column are integers between 1 and 4, indicating the suits of your cards), the cards on the board (a 2 × 5 matrix where the first column represents the numbers of the board cards and the second column represents their suits, and all are 0s for board cards that are not dealt yet), the betting round (integer: 1 = pre-flop, 2 = post-flop, 3 = turn, and 4 = river), the current bet to you (integer), the number of chips you have (integer), the number of chips in the pot (integer), the history of previous bets (vector), the big blind (integer), the number of chips of everyone at the table (vector), your seat number at the table (integer), the seat number of the dealer (integer), and the number of tables remaining in the tournament (integer).

Further details on these variables and functions are given in the *holdem* help files; see for instance *help(gravity)* or *help(tourn1)*.

Another example function called *timemachine* in the *holdem* package decides to go all-in with probability 75% if its total number of chips is less than three times the big blind. The way *timemachine* approximates a 75% probability is by generating a uniform random variable, x, between 0 and 1, and evaluating whether x is less than 0.75.

Figures 8.2 and 8.3 show screenshots of R functions in *holdem* for running a tournament between students' functions; see the function *tourn1()* for details.

As noted in the beginning of this chapter, in addition to running tournaments, R can also be used to calculate or approximate probabilities related to Texas Hold'em. In Example 8.3, a probability is calculated exactly using R. In Example 8.4, R is used to approximate a probability that would be very difficult to compute analytically.

Blinds are 15 and 30		1 table left
D		
Xena	Ursula	Vera
(2950)	(980)	(3070)
Q 9	9 5	K 9
BETS:	BETS:	BETS:
2950	15	2960
2	T	A

FIGURE 8.2 Screenshot of a hand between three remaining players in a tournament. The *D* above *Xena* indicates that Xena is the dealer for the hand. The numbers in parentheses are the players' chip counts at the beginning of the hand. The numbers and letters below their names are their cards; different colors represent the different suits. The numbers and letters at the bottom represent the flop ($T = 10$).

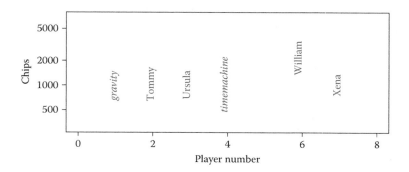

FIGURE 8.3 Sample screenshot showing chip counts in a tournament between students' functions. Note that the *y*-axis is on logarithmic scale.

Example 8.3

In one interesting hand from Season 2 of *High Stakes Poker*, Corey Zeidman (9♥ 9♣) called $800, Doyle Brunson (Q♠ 10♠) raised to $6200, and Eli Elezra (10♥ 10♦), Daniel Negreanu (K♠ J♠), and Zeidman all called. Use *R* to calculate the probability that after the three flop cards are dealt, Elezra would be ahead, in terms of the best *current* five-card hand, of his three opponents. (This would be a feasible but very difficult problem to attack analytically. While many websites and programs report the probability of one hand beating others at showdown, sites or software that allow users to compute probabilities of one hand leading on the flop are non-existent or elusive.)

Answer With eight cards belonging to the four players removed from the deck, there are $C(44,3) = 13,244$ possible flop combinations and all are equally likely to occur. One can use *R* to imitate dealing each of these flops and seeing whether Elezra is leading on the flop for each. After loading the functions in *holdem*, one may use the code below to solve this problem. The first few lines set up the loop of 13,244 iterations, which involves taking out the eight cards in the players' hands and dealing the others. (According to the correspondence used in the function *switch2* between the integers 1 through 52 and the 52 cards in the deck, $1 = 2\clubsuit$, $2 = 3\clubsuit$, etc., so $8 = 9\clubsuit$, $22 = 10\diamondsuit$, $34 = 9\heartsuit$, $35 = 10\heartsuit$, $48 = 10\spadesuit$, $49 = J\spadesuit$, $50 = Q\spadesuit$, and $51 = K\spadesuit$.)

```
n = 13244
result = rep(0,n)
a1 = c(8,22,34,35,48,49,50,51)
a2 = c(1:52)[-a1]
i = 0
for(i1 in 1:42){
for(i2 in ((i1+1):43)){
  for(i3 in ((i2+1):44)){
    flop1 = c(a2[i1],a2[i2],a2[i3])
    flop2 = switch2(flop1)
    b1 = handeval(c(10,10,flop2$num),
      c(3,2,flop2$st))
    b2 = handeval(c(9,9,flop2$num),
      c(3,1,flop2$st))
    b3 = handeval(c(12,10,flop2$num),
      c(4,4,flop2$st))
    b4 = handeval(c(13,11,flop2$num),
      c(4,4,flop2$st))
    i = i+1
    if(b1 > max(b2,b3,b4)) result[i]
      = 1
}}}
sum(result > 0.5)
```

This code, which may take a minute or two to run, loops through all 13,244 possible flops and finds that Elezra is ahead on 5785 of them. Thus the desired probability is $5785/13{,}244 \sim 43.68\%$.

Incidentally, in the actual hand, the flop was $6\diamondsuit$ $9\diamondsuit$ $4\heartsuit$, Zeidman went all-in for \$41,700, Elezra called, and Zeidman won after the turn and river were the uneventful $2\spadesuit$ and $2\diamondsuit$. Note that the answer $5785/13{,}244$ in Example 8.3 is an exact solution, rather than an approximation. The example below illustrates how R can also be used to obtain rapid approximations to complex problems.

Of course, similar types of simulations can be used to approximate probabilities in other (non-poker) applications as well.

Example 8.4

Daniel Negreanu has been on the losing side of some very unfortunate situations where both he and his opponent had extremely powerful hands. For instance, in one hand against Gus Hansen on *High Stakes Poker*, Negreanu had 6♠ 6♥, Hansen had 5♦ 5♣, and the board came 9♣ 6♦ 5♥ 5♠ 8♠. On another hand, against Erick Lindgren, Negreanu had 10♥ 9♥, Lindgren had 8♠ 8♣, and the board came Q♣ 8♥ J♦ 8♦ A♥. Such hands are sometimes called *coolers*. If, for simplicity, we assume that two players are both all-in heads up before their cards are dealt and we define a *cooler* as any hand in which the two players both have straights or better, then what is the probability of a cooler? Perform 100,000 simulations in *R* to approximate the answer, and find a 95% confidence interval for your estimate.

Answer After loading the *holdem* package, the code below may be used to approximate a solution to this problem.

```
n = 100000
result = rep(0,n)
for(i in 1:n){
  x1 = deal1(2)
  b1 = handeval(c(x1$plnum1[1,],x1$br
    dnum1),
   c(x1$plsuit1[1,],x1$brdsuit1))
  b2 = handeval(c(x1$plnum1[2,],x1$br
    dnum1),
   c(x1$plsuit1[2,],x1$brdsuit1))
  if(min(b1,b2) > 4000000) result[i]
    = 1
}
sum(result>.5)
```

During one run of this code, 2505 of the 100,000 simulated hands were coolers. (Different runs of the same code will yield slightly different results.) Thus, we estimate the probability as 2505/100,000 = 2.505%. Using the central limit theorem, a 95% confidence interval for the true probability of a cooler is thus 2.505% ± 1.96 $\sqrt{(2.505\% \times 97.495\%)} \div \sqrt{100{,}000}$

~ 2.505% ± 0.097%, or (2.408%, 2.602%).

Exercises

8.1 Submit a plot of 100 points uniformly scattered inside the unit circle. One simple way to do this is as follows. In *R* first generate a matrix of all 0s of size 100 × 2, which will ultimately store your 100 points. Then make a loop: generate an *x*-coordinate uniformly scattered between −1 and 1, and independently generate a *y*-coordinate uniformly scattered between −1 and 1. If the point is inside the unit circle, then put the point in your matrix. Repeat until you have generated 100 points. Submit only your plot, not the *R* code. You do not need to have *R* draw the circle around the points; just plot the points inside the circle.

8.2 Write an *R* function that takes as inputs your cards and other variables used in *gravity* and decides whether you go all-in or fold. The function must return a 0 integer if you fold or your number of chips if you go all-in.

8.3 Write an *R* function that takes as inputs your cards and other variables used in *gravity* and that returns an integer indicating the number of chips to wager where 0 means fold.

8.4 Suppose all four players in a game of Texas Hold'em are all-in before their cards are dealt. As in Example 8.4, define a cooler as any hand in which at least two players have straights or better. What is the probability of a cooler? Perform 10,000 simulations in R to approximate the answer and find a 95% confidence interval for this probability.

8.5 Consider the information and assumptions from Exercise 7.8 in which Daniel Negreanu lost approximately $1.7 million in total over the first five seasons of *High Stakes Poker* or about $1360 per hand over ~1250 hands. If a player's long-term mean were actually $0 per hand and the *SD* of his winnings were $30,000 per hand, using the central limit theorem, what would be the probability that this player would have a sample mean of −$1360 or less over the course of 1250 hands? Use the *pnorm* () function in R to calculate your answer.

8.6 Assuming two players are heads up and all-in next hand regardless of their cards, use 1000 simulations to approximate the probability that both of them end up with ace high or worse. Report your answer as a 95% confidence interval for this probability.

8.7 Assuming two players are heads up and all-in next hand regardless of their cards, use 10,000 simulations to approximate the probability that both of them end up with a full house or better. Report your answer as a 95% confidence interval for this probability.

8.8 Suppose a player starts with 100 chips and every minute she either gains 1 chip with probability 0.51 or loses 1 chip with probability 0.49. Use 1000 simulations to approximate the probability that the player has at least 90 chips after 100 minutes. Report your answer as a 95% confidence interval for this probability.

8.9 Suppose a player starts with 100 chips and every minute she either gains 1 chip with probability 0.55 or loses 1 chip with probability 0.45. Use 1000 simulations to approximate the probability that the player has at least 90 chips after 100 minutes. Report your answer as a 95% confidence interval for this probability.

8.10 Does conservative play pay off? Consider the following scenario. Suppose a tournament has three players, and each has 10 chips. Every minute, two of them are chosen, and the other is excluded. Of the two players competing, one of them wins a chip from the other, and each player has a 50% chance of winning the chip from the other player. However, suppose player 3 is more conservative and player 1 is less conservative, so that when choosing two players at random to compete, with probability 20% player 1 is the one excluded, with probability 30% player 2 is the one excluded, and with the remaining 50% probability player 3 is the one excluded. When any player has no chips left, she is eliminated and the other two players continue. First place gets $100, second place gets $50, and third place gets $0. Run 1000 simulations and report 95% confidence intervals for the average winnings per tournament for each player.

Appendix A: Abbreviated Rules of Texas Hold'em

1. Deck

Texas Hold'em uses a standard deck of 52 cards. Each card has a *number* and a *suit*. The four *suits* are clubs (♣), diamonds (♦), hearts (♥), and spades (♠). The *numbers* are 2, 3, 4, 5, 6, 7, 8, 9, 10, jack (J), queen (Q), king (K), and ace (A). Each card is unique within a deck because it combines one of the thirteen numbers and one of the four suits, for example, J♦.

2. Posting Blinds and Dealing Hole Cards

A button is placed before one of the players. The player to the left of the button bets a certain amount called the *small blind*, and the player two seats to the left of the button bets the *big blind*—typically twice the small blind. Sometimes all players must also pay antes into the pot. After these bets are made, each player is dealt two cards facedown from the deck. These are known as *hole cards*. Players may then look at their two hole cards.

3. Initial Betting Round

Section 7 describes betting rounds. On the initial betting round, the small blind is considered the first bet, and the big blind is considered a raise. The first player to act is the player to the left of the big blind, and thus the initial *current bet* is the amount of the big blind.

4. Flop

The dealer deals three cards from the deck face up in the middle of the table. These cards are called the *flop*. After the flop is dealt, another betting round occurs.

5. Turn

The dealer deals one card from the deck face up in the middle of the table. This card is called the *turn*. After the turn is dealt, another betting round occurs.

6. River

The dealer deals one final card from the deck face up in the middle of the table. This card is called the *river*. After the river is dealt, a final betting round occurs.

7. Betting Rounds in General

Each betting round starts with the player to the left of the button and proceeds clockwise. The maximum bet on a betting round is called the *current bet*; initially, on the first betting round, the current bet is the big blind. On subsequent betting rounds, the current bet is initially 0. During each player's turn, he or she may act by calling the current bet (wagering the minimum amount needed to stay in the hand), raising (betting more than the current bet), or folding (discarding the hole cards and forgoing any chance of winning the pot).

If the current bet is 0, the player may bet 0—this is called *checking*. If the betting is on a certain player who has already acted during the betting round and all *other* players have either folded or called the current bet, then the betting round is over (to determine when a betting round is over, checking is considered equivalent to calling a bet of 0; posting the small blind or big blind is not considered acting). If only one player remains without folding, he or she wins the pot, and the hand is over.

8. Determining the Winner

After the final betting round, among players who have not folded, the player with the best hand wins the pot. Players can use any combination of five of the seven cards (five community cards plus two hole cards) to form their best possible five-card poker hands. The ordering of the poker hands is described in Section 9. After each hand, the button rotates clockwise to the next player.

9. Hand Rankings

Five-card poker hands are ranked here from best to worst:

Straight flush (4♦ 5♦ 6♦ 7♦ 8♦)
Four of a kind (J♣ J♦ J♥ J♠ 3♣)
Full house (5♣ 5♦ 5♠ 7♥ 7♣)
Flush (A♥ J♥ 10♥ 5♥ 2♥)
Straight (Q♦ J♥ 10♠ 9♠ 8♣)
Three of a kind (10♠ 10♣ 10♦ K♣ 4♦)
Two pairs (5♥ 5♦ 3♣ 3♥ A♠)
One pair (K♥ K♦ 9♠ 7♣ 3♥)
Nothing (A♠ K♠ Q♠ J♠ 9♥)

There are other rules covering such issues as minimal bet sizes, all-in wagers, and split pots. For example, the amount of any raise must be at least as large as the previous bet. In addition, most casinos impose rules on etiquette and behavior.

Appendix B: Glossary of Poker Terms

All-in: All one's chips.

Ante: Forced bet that everyone at the table must post before each hand is dealt.

Big blind: Bet that must be posted by the player two seats to the left of the dealer before the hole cards are even dealt or the player who must post this bet. The amount of the big blind is typically double the amount of the small blind.

Blind: Without looking at cards; also, the bet a player must post before the cards are even dealt.

Bluff: Bet with a weak hand, with the intention of winning the pot by having all opponents fold.

Board: Five community cards consisting of the flop, turn, and river; or the portion of these community cards currently exposed.

Button: Disk placed in front of one player indicating that he or she bets last on the betting rounds; also, the player in this seat.

Call: Match bet made by previous player.

Check: Bet zero chips.

Community cards: Five cards consisting of the flop, turn, and river that are used with hole cards to form the best five-card poker hand.

Cooler: Two or more players having extremely powerful hands at the same time.

Cut-off: Seat to the right of the dealer; also the player in that seat.

Dealer: Person dealing the cards. Traditionally this was the player with the button at his or her seat. Modern card rooms and casinos employ a dealer at each table who does not play, but merely shuffles the cards, deals, and awards the pot to the winner at the end of each hand.

Face card: King, queen, or jack.

Flop: First three community cards.

Flush: Five cards of same suit.

Flush draw: Needing only one more card of a suit to make a flush. For example, if you have Q♠ 10♥ and the flop is 7♠ A♠ 3♠, then you have a flush draw.

Hand: Single game of Texas Hold'em; within the game, a hand can mean a player's hole cards or best five-card combination using hole cards and community cards.

Heads up: Between only two players; one against one.

High pocket pair: Pocket pair of 10s, jacks, queens, kings, or aces.

Hole cards: Two cards dealt to each player face down at the beginning of a hand.

Led out: Bet.

Limp: Call the big blind.

Muck: Fold.

Nuts: Best current hand possible given the board. For example, if the board is K♠ 10♦ 9♦ 7♥, and your hole cards are Q♠ J♥, then you have the nuts. Note that someone else could have Q♦ J♦ and thus have a better chance of winning, but you nevertheless are said to have the nuts with Q♠ J♥ because you have the best hand possible at the moment.

Open-ended straight draw: Needing one of two possible numbers to complete a straight; see straight draw.

Out: A board card that would give you the lead over your opponent.

Pair: Two cards of the same number.

Playing the board: Exclusively using the five board cards, rather than either of your two hole cards, to form your best five-card poker hand.

Pocket aces: Two hole cards that are both aces.

Pocket pair: Two hole cards of the same number (e.g., Q♣ Q♥).

Post: Bet (used in reference to forced bets such as blinds or antes).

Raise: Bet more than the previous bettor.

Rake: Amount taken from the pot and awarded to the casino.

Re-raise: Bet more than the previous raiser.

River: Fifth and final community card.

Royal flush: A, K, Q, J, and 10 of same suit; the highest possible straight flush and thus the highest possible hand.

Runner–runner flush draw: Needing two more board cards of your suit in order to make a flush.

Running it twice: In cash games, players may opt to have the dealer deal the remaining board cards twice (without reshuffling the cards into the deck between the two deals), with each deal worth half the pot.

Semi-bluffing: Betting when one is currently behind but has outs, and thus can have a decent chance to win even if opponents call.

Small blind: Bet that must be posted by the player sitting to the left of the dealer before the hole cards are dealt; also, the player who must post this bet. The amount of the small blind is typically one-half the amount of the big blind.

Split pot: Pot divided evenly between two or more players whose best five-card hands are equivalent.

Straight: Five cards with numbers in sequence, such as 3♥ 4♣ 5♠ 6♣ 7♣, or 8♦ 9♥ 10♦ J♣ Q♠. In Texas Hold'em, aces can be high or low, so both A2345 and 10JQKA are straights; JQKA2, QKA23, and KA234 are not considered straights.

Straight draw: Needing only one more card to make a straight. For instance, if you have 6♠ 5♦ and the flop comes 4♠ 7♥ K♣, then you have a straight draw. This specific case is called an *open-ended straight draw*, because two numbers (3 or 8) will make a straight for you if either appears on the next card. If the flop came 4♠ 8♥ K♣, then only a 7 would make a straight. When only one possible number will make a straight for you, you have a gunshot straight draw.

Straight flush: Five cards of the same suit in sequence, such as 3♥ 4♥ 5♥ 6♥ 7♥. As with straights, both A♦ 2♦ 3♦ 4♦ 5♦ and 10♣ J♣ Q♣ K♣ A♣ are straight flushes; J♣ Q♣ K♣ A♣ 2♣, Q♦ K♦ A♦ 2♦ 3♦, and K♠ A♠ 2♠ 3♠ 4♠ are flushes but not straight flushes.

Suited connectors: Hole cards of the same suit and only one number apart, such as 6♦ 7♦ or Q♠ J♠.

Three-bet: Re-raise.

Turn: Fourth community card.

Unbreakable nuts: Not only the currently best five-card poker hand possible given the current board but also the best five-card poker hand possible regardless of what future board cards will come. For instance, if you have 7♦ 8♦ and the flop is 4♦ 5♦ 6♦, then you have the unbreakable nuts, but if the flop is 5♦ 6♦ 9♥, you do not because the turn and river could be the 10♦ and J♦ and someone could have the Q♦ K♦ and beat you.

Under the gun: First player to act in a hand; player to the left of the big blind.

Value bet: Bet intending an opponent with an inferior hand to call.

VPIP: The percentage of hands one voluntarily puts money in pot; excludes the hands in which one pays only the blinds since these are forced actions.

WSOP: World Series of Poker.

Appendix C: Solutions to Selected Odd-Numbered Exercises

1.1 It is 3.98%, because exactly one of these events must happen, and the other probabilities sum to 96.02%.

1.3 The fact that only one of the events can occur ensures that $A_1, A_2, \ldots, A_n$ are mutually exclusive. Because one of the events must occur, $P(A_1 \text{ or } \ldots \text{ or } A_n) = 1$, so by axiom 3, $P(A_1) + \ldots + P(A_n) = 1$. The events are equally likely, so $P(A_1) = P(A_2) = \ldots = P(A_n)$, and thus $n\,P(A_1) = 1$, so $P(A_1) = 1/n$.

2.3 $[3 \times C(4,2) + 4 \times 4] \div C(52,2) = 34/1326 = 1$ in 39, or ~2.56%.

2.5 No matter which pocket pair you have, the $C(50,3)$ possible flops are all equally likely, and 48 give you four of a kind (e.g., if you have 7♣ 7♥, you need to flop 7♦ 7♠ x and there are 48 choices for x). So $P(\text{flop four of a kind} \mid \text{pocket pair}) = 48/C(50,3)$ or about 1 in 408.33.

2.7 Your flush may be one of four possible suits. One of the cards must be an ace. The number of possibilities for the others is $C(12,4)$. So P (flop ace high flush) $= 4 \times C(12,4)/C(52,5) = 1$ in 1312.606.

2.9 Be careful on this problem not to double-count combinations such as (4♣ 4♠ 9♣ 9♥ Q♣) and (9♣ 9♥ 4♣ 4♠ Q♣). There are $C(13,2)$ possible choices for the *numbers* on the two pairs. For any such choices, there are $C(4,2)$ possibilities for the suits on the higher pair and $C(4,2)$ possibilities for the suits on the lower pair. Finally, for any choices of the above, 44 possibilities remain for the fifth card. So the probability of flopping two pairs is $C(13,2) \times C(4,2) \times C(4,2) \times 44 \div C(52,5) \sim 4.75\%$ or about 1 in 21.035.

2.11 Letting a = any remaining card other than a 6 or 10,

b = any remaining card other than a 6, 7, or Q,

c = any remaining card other than 6, 7, 8, 9, 10, J, or Q, and

d = any remaining card other than 6, 7, 10, or J.

P(Gold wins) = P(turn and river are 66 or 77 or 1010 or JJ or 6a or 7b or 10c or Jd)

$= [C(4,2) + C(2,2) + C(3,2) + C(4,2) + 4 \times 38 + 2 \times 37 + 3 \times 24 + 4 \times 32] \div C(45,2)$

$= 442 \div 990 \sim 44.65\%$.

2.13 $[3 \times C(4,2)] \div C(52,2) = 18/1326 \sim 1.36\%$.

2.15 In order to make a royal flush, the board must contain the Q♣, J♣, and 10♣ along with any two other cards; there are $C(47,2)$ possible choices for these two other cards. So, the probability of making a royal flush is $C(47,2) \div C(50,5) = 1$ in 1960 $\sim 0.051\%$.

2.17 P(KKK or JJJ or AQ10 or Q109 or KKJ but not AQ10 or Q109 of the same suit)

$= [C(3,3) + C(3,3) + 4 \times 4 \times 4 + 4 \times 4 \times 4 + C(3,2) \times 3 - 4 - 4] \div C(50,3)$

$= 131/19,600 \sim 0.668\%$ or about 1 in 149.6.

2.19 Galfond has 0 probability of winning the hand. A split pot can only happen if the turn and river are both 9s, which has a probability of $1/C(45,2) = 1/990 \sim 0.101\%$.

3.1 (a) $O_A = p/(1-p)$, so $(1-p)O_A = p$, so $O_A - pO_A = p$, so $O_A = pO_A + p = p(O_A + 1)$. Thus, $p = O_A \div (O_A + 1)$.

(b) Using part (a), $p = 1/10 \div 11/10 = 1/11$.

(c) $5 = P(A^c)/P(A) = (1-p)/p$, so $5p = 1 - p$, so $6p = 1$. Thus, $p = 1/6$.

3.3 P(both cards are clubs | both cards are black)
= P(both cards are clubs and black)/P(both cards are black)
= P(both cards are clubs)/P(both cards are black) [because if they are clubs, then they must be black]
= $[C(13,2) \div C(52,2]/[C(26,2) \div C(52,2)]$
= $C(13,2)/C(26,2)$
= 24.0%, or about 1 in 4.17.

3.5 No. $P(AB) = P$ (your cards are aces and your cards are black)
   = $P(A\clubsuit\ A\spadesuit)$
   = $1/C(52,2)$
   = 1/1326 ~0.0754%.

$P(A) = P(AA) = C(4,2)/C(52,2) = 6/1326$ ~0.452%.
$P(B) = P$(both cards are black) = $C(26,2)/C(52,2)$
   = 25/102 ~24.51%.
So, $P(A)P(B) = 6/1326 \times 25/102 = 150/135252$

~1 in 901.68, or 0.111%.

Thus, $P(AB) \neq P(A)P(B)$.

3.9 Let B_1, B_2, and B_3 represent the events that she has AA, KK, or AK, respectively. Let A represent the event that she displays a tell. By assumption, $P(A \mid B_1) = P(A \mid B_2) = 100\%$, and $P(A \mid B_3) = 50\%$. $P(B_1) = P(B_2) = C(4,2)/C(52,2) = 1/221$, and $P(B_3) = 4 \times 4/C(52,2) = 16/1326$.
By Bayes' rule,
$P(B_1 \mid A) = P(A \mid B_1)\, P(B_1) \div [P(A \mid B_1)\, P(B_1) + P(A \mid B_2)$
$\qquad\qquad P(B_2) + P(A \mid B_3)\, P(B_3)]$
$\qquad\quad = 100\% \times 1/221 \div [100\% \times 1/221 + 100\% \times$
$\qquad\qquad 1/221 + 50\% \times 16/1326]$
$\qquad\quad = 30\%$.

4.3 (a) If $A\diamondsuit\ J\heartsuit$ folds, $P(K\clubsuit\ K\spadesuit$ wins) ~80.93%, and $P(K\clubsuit\ K\spadesuit$ ties) ~0.41%, so your expected number of chips after the hand ~\$200 (80.93%) + \$100 (0.41%) + \$0 (18.66%) = \$162.27.

(b) If A♦ J♥ calls, P(K♣ K♠ wins) ~ 57.97% and P(K♣ K♠ ties) ~0.30%, so your expected number of chips after the hand ~\$300 (57.97%) + \$100 (0.30%) + \$0 (41.73%) = \$174.21. Thus, your expected number of chips is higher if the player with A♦ J♥ calls.

4.5 (a) There are nine spades left, and he could make a flush if the turn and river are both spades or if the turn and river contain exactly one spade. There are $C(9,2)$ combinations involving two spades on the turn and river, and 9×36 involving one spade, so the probability of at least one spade on the turn and river is $[C(9,2) + 9 \times 36]/C(45,2) = 360/990$ ~36.36%.

(b) If Negreanu makes a flush, Ly can still win the hand by making a full house or four of a kind. This can occur if the turn and river contain the 7♠ and another 7, or the 7♠ and K♣, or the K♠ and K♣, or the K♠ and a 2, 7, or 8. Counting combinations, the probability associated with one of these events occurring is $[1 \times 2 + 1 + 1 + 1 \times 9]/C(45,2) = 13/990$ ~1.31%.

(c) Negreanu can make a straight with the turn and river containing a 9 and a 10 with no spades; the probability of this is $(3 \times 3)/C(45,2) = 9/990$ ~0.91%.

(d) Negreanu can win the hand without a flush or straight if the turn or river contains both aces or both jacks, or an ace and x, where x is any non-king, non-spade, and non-ace. The probability of this event is $[C(3,2) + C(3,2) + 3 \times 32]/C(45,2) = 102/990$ ~10.30%.

(e) $360/990 - 13/990 + 9/990 + 102/990 = 458/990$ ~46.26%.

(f) \$18,000/(\$11,000 + \$11,000 + \$18,000 + \$18,000) = \$18,000/\$58,000 ~31.03% < 46.26%, so yes, Negreanu would have been correct to call if his goal were to maximize his expected number of chips after the hand.

4.13 Esfandiari can win if the turn, river, or both contain a heart, 5, or 9, provided Greenstein does not make four of a kind or a full house. Esfandiari also wins if the turn and river are the 9♥ and 10♥ or if the 5♥ comes and Greenstein makes four of a kind or a full house. We can write the combinations where Esfandiari wins as follows: $(a, b), (a, c), (5n, d), (9n, e)$, $(9♥, 10♥)$, and $(5♥, f)$, where a and b represent any hearts other than 10♥; c is any card that is neither a heart, 10, 6, or 4 nor the same number as a; n signifies any suit besides hearts; d is any non-heart♥ card other than any 10, 6, 4, or 5; e is any non-heart card♥ other than any 10, 6, 4, 5, or 9; and f is any 10, 6, 4, or 5. Note that straights, flushes, and straight flushes are the only possible ways Esfandiari can win the hand. Thus, if he calls, the probability of Esfandiari winning the hand is $[C(8,2) + 8 \times 27 + 3 \times 27 + 3 \times 24 + 1 + 1 \times 10]/C(45,2) = 408/990 \sim 41.21\%$. If Esfandiari calls the \$181,200 bet, the pot size will be \$800 + \$2500 × 5 + \$6200 + \$106,000 × 2 + \$181,200 × 2 = \$593,700. Since \$181,200 ÷ \$593,900 ~30.51% < 41.21%, Esfandiari would have been correct to call (as he did), if his goal were to maximize his expected number of chips after the hand.

4.15 (a) $\phi_Y(t) = E[e^{tY}] = E[\exp\{t(3X + 7)\}] = E[\exp(3tX)\exp(7t)] = \phi_X(3t)\exp(7t)$.

(b) $\phi_Y(2) = \phi_X(6)\exp(14) = 0.001 \times \exp(14) \sim 1202.604$.

5.1 The probability of being dealt (10, 2) is $4 \times 4/C(52,2) = 16/1326$. Given that you have (10, 2), you make a full house if the board comes *aaabb*, 10 10 *aaa*, 10 10 *aab*, 2 2 *aaa*, 2 2 *aab*, 10 10 2 *aa*, 10 10 2 *ab*, 10 2 2 *aa*, 10 10 2 2 *aa*, or 10 2 2 *ab*, 10 10 2 2 *ab*, where a and b can be any numbers aside from 10 and 2 such that $a \neq b$. Thus given that you have (10, 2), the probability of your making a full house is $[(11 \times C(4,3) \times 10 \times C(4,2)) + (C(3,2) \times 11 \times C(4,3) + C(3,2) \times 11 \times C(4,2) \times 10 \times 4) + (C(3,2) \times 11 \times C(4,3)) + (C(3,2) \times 11 \times C(4,2) \times 10 \times 4) + (C(3,2) \times 3 \times 11 \times C(4,2)) + (C(3,2) \times 3 \times C(11,2) \times 4 \times 4 + (C(3,2) \times 3 \times 11 \times C(4,2)) + (C(3,2) \times C(3,2) \times 11 \times 4) + (C(3,2) \times 3 \times C(11,2) \times 4 \times 4)] \div C(50,5) =$ 36,168/2,118,760. Thus, $P\{$you are dealt (10, 2) and make a full house$\} = P\{$you are dealt (10, 2)$\} \times P\{$you make a full house | you are dealt (10, 2)$\} = 16/1326 \times$ 36,168/2,118,760 ~ 0.0206%, or 1 in 4854.906. $P(X$ is at least 2$) = 1 - P(X = 0) - P(X = 1) = 1 - C(100,0)$ $(0.0206\%)^0$ $(99.9794\%)^{100} - C(100,1) (0.0206\%)^1$ $(99.9794\%)^{99}$ ~ 0.0207%.

5.3 $P(X = k \mid X > c) = P(X = k$ and $X > c)/P(X > c) = P(X = k)/P(X > c) = q^{k-1}p/qc = q^{k-c-1}p = f(k - c)$.

5.5 Let $p = P(AA) = C(4,2)/C(52,2) = 1/221$.
Let $q = 1 - p = 220/221$.
$E(X) = 1/p = 221$.
$V(X) = q/p^2 = 220 \times 221 = 48,620$, so $SD(X) = \sqrt{48,620}$ ~220.50.
Let $r = P($high pocket pair$) = 5 \times C(4,2)/C(52,2) = 5/221$. Let $s = 1 - r = 216/221$.
$E(Y) = 1/r = 221/5 = 44.20$.
$V(Y) = s/r^2 = 216/5 \times 44.20 = 1909.44$, so $SD(Y) = \sqrt{1909.44}$ ~43.70.

The expectation and *SD* of the waiting time for a high pocket pair are both much shorter than the expectation and *SD* of the waiting time for pocket aces.

5.7 This can be calculated using the moment-generating function,
$\phi_X(t) = E\{\exp(tX)\} = \Sigma\exp(tk) \, q^{k-1}p = p\exp(t)$
$\Sigma[\exp(t)q]^k$
$= p\exp(t)/[1 - q\exp(t)]$.
$\phi'(t) = p\exp(t)/[1 - q\exp(t)] + pq\exp(2t)[1 - q\exp(t)]^{-2}$.
$\phi''(t) = \phi'(t) + 2pq\exp(2t)[1 - q\exp(t)]^{-2} + 2pq^2\exp(3t)$
$[1 - q\exp(t)]^{-3}$,

so $E(X^2) = \phi''(0)$
$$= \phi'(0) + 2pq(1-q)^{-2} + 2pq^2(1-q)^{-3}$$
$$= 1/p + 2pq(1-q)^{-2} + 2pq^2(1-q)^{-3}$$
$$= 1/p + 2q/p + 2q^2/p^2.$$
Thus, $\text{Var}(X) = E(X^2) - [E(X)]^2$
$$= 1/p + 2q/p + 2q^2/p^2 - 1/p^2$$
$$= p^{-2}(p + 2pq + 2q^2 - 1)$$
$$= p^{-2}[p + 2p(1-p) + 2(1-p)^2 - 1]$$
$$= p^{-2}(p + 2p - 2p^2 + 2 - 4p + 2p^2 - 1)$$
$$= p^{-2}(1-p)$$
$$= q/p^2.$$

6.3 Let $Y = F(X)$, and for c in $(0,1)$, let $d = F^{-1}(c) = \inf\{z : F(z) \geq c\}$. $P\{Y \leq c\} = P\{F(X) \leq c\} = P\{X \leq F^{-1}(c)\} = P\{X \leq d\} = F(d) = c$.

6.5 $P(X > \$50,000) \div P(X > \$25,000) = (1/2)^b = 1/8$, so $b = 3$ and $P(X > \$100,000) = 1/8 \ (10\%) = 1.25\%$.

6.7 Exponential. $P(Y > c) = P(\log X > c) = P\{X > \exp(c)\} = (a/\exp(c))^b = \exp(-bc)$, since $a = 1$. Thus Y is exponential with parameter $\lambda = b$.

6.9 Let $F(c)$ be the cdf of Z. For $0 < c < 1$, $F(c) = P(Z \leq c) = P(XY \leq c) = P(X \leq c/2 \text{ and } Y = 2) + P(X \leq c \text{ and } Y = 2) = P(X \leq c) \ P(Y = 1) + P(X \leq c/2)P(Y = 2)$ [since X and Y are independent] $= (c)(1/3) + (c/2)(2/3) = c/3 + c/3 = 2c/3$. Thus $f(c) = F'(c) = 2/3$.

For $1 < c < 2$, $F(c) = P(X \leq c \text{ and } Y = 1) + P(X \leq c \text{ and } Y = 2) = P(X \leq c) \ P(Y = 1) + P(X \leq c/2)P(Y = 2) = (1)(1/3) + c/3 = (c + 1)/3$, so $f(c) = F'(c) = 1/3$.

$$E(Z) = \int_0^1 c(2/3)\,dc + \int_1^2 c(1/3)\,dc = 1/3 + 1/2 = 5/6.$$

$$E(Z^2) = \int_0^1 c^2(2/3)\,dc + \int_1^2 c^2(1/3)\,dc = 2/9 + 7/9 = 1.$$

So, $V(Z) = E(Z^2) - [E(Z)]^2 = 1 - 25/36 = 11/36$, and $SD(Z) = \sqrt{(11/36)} \sim 0.5528$.

7.3 (a) $P(\text{winning the tournament}) = P(\text{double up eight times}) = 56\%^8 \sim 0.967\%$, or about 1 in 103.4.
   (b) The expected profit per tournament is (\$255,000) $(56\%^8) + (-\$1000)(1 - 56\%^8) \sim \$1475.52$.

7.7 Among many possible answers, one is $x_i = \sqrt{i} - \sqrt{(i-1)}$ for $i = 1, 2, \ldots$.

Thus, $\Sigma_{i=1}{}^n x_i = x_1 + x_2 + \ldots + x_n = (\sqrt{1} - \sqrt{0}) + (\sqrt{2} - \sqrt{1}) + \ldots + [\sqrt{n} - \sqrt{(n-1)}] = \sqrt{n} \to \infty$, and $\Sigma_{i=1}{}^n x_i/n = \sqrt{n}/n = 1/\sqrt{n} \to 0$. While the law of large numbers implies that the *average* of independent observations with mean 0 will generally converge to 0, the *sum* may not only fail to converge to 0 but could in fact diverge to infinity.

7.9 Since X and Y are independent, $P(Y = k \mid X = j) = P(Y = k)$ for any values of j and k. Thus, for each value of j,

$$E(Y \mid X = j) = \Sigma_k k\, P(Y = k \mid X = j) = \Sigma_k k\, P(Y = k)$$
$$= E(Y).$$

7.13 By Theorem 7.6.8, the probability that you eventually win the tournament is $(1 - r^k)/(1 - r^n) = (1 - q^k/p^k)/(1 - q^n/p^n)$. Similarly, applying Theorem 7.6.7 to your opponent, who has $n - k$ chips and probability $q = 1 - p$ of gaining a chip on each hand, implies that your opponent's probability of winning the tournament is $(1 - p^{n-k}/q^{n-k})/(1 - p^n/q^n)$. Thus, the sum of these two probabilities is $(1 - q^k/p^k)/(1 - q^n/p^n) + (1 - p^{n-k}/q^{n-k})/(1 - p^n/q^n)$. Multiplying the numerator and denominator of the first term by p^n and the second term by $-q^n$, we find the sum of the two probabilities

$= (p^n - p^{n-k}q^k)/(p^n - q^n) + (p^{n-k}q^k - q^n)/(p^n - q^n) = (p^n - p^{n-k}q^k + p^{n-k}q^k - q^n)/(p^n - q^n) = (p^n - q^n)/(p^n - q^n) = 1$.

Since the probability that you or your opponent will eventually win the tournament is 1, the probability that neither of you wins is therefore 0.

7.15 $2^9 = 512$, so you need to double up nine times to win the tournament. Thus, $p^9 = 5\%$, so $\log p = \log(5\%)/9$, and $p = \exp\{\log(5\%)/9\} \sim 71.69\%$.

References and Suggested Reading

Bayes, T. and Price, R. 1763. An essay towards solving a problem in the doctrine of chance. By the late Mr. Bayes, communicated by Mr. Price, in a letter to John Canton, M.A. and F.R.S. *Philosophical Transactions of the Royal Society of London.* 53: 370–418.

Bertsekas, D. and Tsitsiklas, J. 2008. *Introduction to Probability*, 2nd ed. Athena Scientific, Nashua, NH.

Billingsley, P. 1990. *Probability and Measure*, 2nd ed. Wiley, New York.

Blackwell, D. and Girshick, M.A. 1954. *Theory of Games and Statistical Decisions*. Dover, New York.

Borel, É. 1938. *Traité du Calcul des Probabilités et ses Applications*, Fasc. 2, Vol. 4, Applications aux jeux des hazard. Gauthier-Villars, Paris.

Breiman, L. 1961. Optimal gambling systems for favorable games. *Fourth Berkeley Symposium on Mathematical Statistics and Probability.* 1: 65–78.

Brunson, D. and Addington, C. 2002. *Doyle Brunson's Super System: A Course in Power Poker*, 3rd ed. Cardoza, New York.

Brunson, D. and Addington, C. 2005. *Doyle Brunson's Super System 2: A Course in Power Poker*. Cardoza, New York.

Caro, M. 2003. *Caro's Book of Poker Tells*. Cardoza, New York.

Chen, W. and Ankenman, J. 2006. *The Mathematics of Poker*. ConJelCo, Pittsburgh, PA.

Chung, K.L. 1974. *A Course in Probability Theory*, 2nd ed. Academic Press, New York.

Chung, K.L. and Aitsahlia, F. 2003. *Elementary Probability Theory with Stochastic Processes and an Introduction to Mathematical Finance*, 4th ed. Springer, New York.

Dedonno, M.A. and Detterman, D.K. 2008. Poker is a skill. *Gaming Law Rev.* 12: 31–36.

Durrett, R. 2010. *Probability: Theory and Examples*, 4th ed. Cambridge University Press, New York.

Feller, W. 1967. *Introduction to Probability Theory and Its Applications*, 3rd ed., Vol. 1. Wiley, New York.

Feller, W. 1966. *Introduction to Probability Theory and Its Applications*, 3rd ed., Vol. 2. Wiley, New York.

Ferguson, C. and Ferguson, T. 2003. On the Borel and von Neumann poker models. *Game Theory Appl.* 9: 17–32.

Ferguson, C. and Ferguson, T. 2007. The endgame in poker. In *Optimal Play: Mathematical Studies of Games and Gambling*. Institute for Study of Gambling and Commercial Gaming, Reno, pp. 79–106.

Ferguson, C., Ferguson, T., and Gawargy, C. 2007. Uniform (0,1) two-person poker models. *Game Theory Appl.* 12: 17–37.

Goldberg, S. 1986. *Probability: An Introduction*. Dover, New York.

Gordon, P. 2006. *Phil Gordon's Little Blue Book: More Lessons and Hand Analysis in No Limit Texas Hold'em*. Simon & Schuster, New York.

Grinstead, C.M. and Snell, J.L. 1997. *Introduction to Probability*, 2nd rev. ed. American Mathematical Society, Providence, RI.

Harrington, D. and Robertie, B. 2004. *Harrington on Hold'em: Expert Strategy for No Limit Tournaments*. Vol. 1: *Strategic Play*. Two Plus Two Publishing, Henderson, NV.

Harrington, D. and Robertie, B. 2005. *Harrington on Hold'em: Expert Strategy for No Limit Tournaments*. Vol. 2: *The Endgame*. Two Plus Two Publishing, Henderson, NV.

Harrington, D. and Robertie, B. 2006. *Harrington on Hold'em: Expert Strategy for No Limit Tournaments*. Vol. 3: *The Workbook*. Two Plus Two Publishing, Henderson, NV.

Hellmuth, P. 2003. *Play Poker Like the Pros*. HarperCollins, New York.

Karr, A.F. 1993. *Probability*. Springer, New York.

Kelly, J. L., Jr. 1956. A new interpretation of information rate. *Bell System Tech. J.* 35: 917–926.

Kim, M.S. 2005. *Gambler's Ruin in Many Dimensions and Optimal Strategy in Repeated Multi-Player Games with Application to Poker*. Master's Thesis, University of California, Los Angeles, pp. 1–32.

Malmuth, M. 2004. *Poker Essays*, Vol. 3. Two Plus Two Publishing, Henderson, NV.

Pitman, J. 1993. *Probability*. Springer, New York.

Ross, S. 2009. *A First Course in Probability*, 8th ed. Prentice Hall, New York.

Schoenberg, F.P. and Patel, R.D. 2012. Comparison of Pareto and tapered Pareto distributions for environmental phenomena. *Eur Phys J* 205: 159–166.

Sklansky, D. 1989. *Theory of Poker*, 3rd ed. Two Plus Two Publishing, Henderson, NV.

Sklansky, D. and Miller, E. 2006. *No Limit Hold'em: Theory and Practice*. Two Plus Two Publishing, Henderson, NV.

Thorp, E.O. 1966. *Beat the Dealer: A Winning Strategy for the Game of Twenty-One*, 2nd ed. Blaisdell, New York.

Thorp, E.O. and Kassouf, S.T. 1967. *Beat the Market: A Scientific Stock Market System*. Random House, New York.

Varadhan, S.R.S. 2001. *Probability Theory*. Courant Lecture Notes in Mathematics, Vol. 7. American Mathematical Society, Providence, RI.

von Neumann, J. and Morgenstern, O. 1944. *Theory of Games and Economic Behavior*. Princeton University Press, Princeton, NJ.

Index

A

Addition rule,
    for mutually exclusive events, 4
    general, 7
Admissibility, 174
Affleck, Matt, 18
Aggressive play, 92
Ahmar, Wasim, 135–136
Ankenman, Jerrod, 167–168, 222, 269
Antonius, Patrik, 217, 219
Arcsine law, 226
Area and probability, 6, 7, 58, 61, 158, 160
Axioms of probability, 4

B

Ballot theorem, 223–224
Bardah, Ronnie, 154
Bayesian view of probability, 2, 13
Bayes's rule, 71–75, 184–187
Beckley, Josh, 54, 82, 134
Bennet, Rick, 234
Benyamine, David, 52
Bernoulli, Jacob, 138, 200
Bernoulli random variable, 84–86,
    137–141, 202, 214, 220
Billingsley, Patrick, 125, 127, 269
Binger, Michael, 1, 8, 10–13, 33, 35,
    98–101
Binomial random variable, 137, 140–142,
    148, 149, 151, 153, 154, 211
Blackwell, David, 169, 269
Bluffing, 73, 81, 93, 102–104, 117, 119,
    133, 134, 136, 147, 150,
    169–171
Blumenfield, Niel, 52, 54, 82, 135–136
Bonomo, Justin, 80
Boole's inequality, 13
Booth, Brad, 73–74
Borel, Emil, 168, 269–270
Brand, Chris, 55, 79–80
Breiman, Leo, 228, 269
Brunson, Doyle, 43, 146, 152, 246, 269

Brunson, Todd, 68
Bubble, 92
Butteroni, Federico, 55, 79–80

C

Cannuli, Tom, 82, 135
Cardano, Gerolamo, 200
Caro, Mike, 78, 269
CDF, see Cumulative distribution
    function
Central limit theorem, 191, 208–216,
    240, 249, 250
Chebyshev's inequality, 123–125,
    187, 201
Chen, William, 90–91, 95, 167–168,
    222, 269
Childs, Lee, 19–20
Clinger, Randy, 54
Cloud, Mike, 63, 108–109
Coin flip, 65, 119, 227, 233
Combinations, 11, 19–20, 26–49, 62–64
Complement, 4, 9, 37
Computation using R, 239–249
Conditional density, 184
Conditional expectation, 197–199
Confidence intervals, 215–220
Continuous posterior distribution,
    see Posterior distribution
Continuous prior distribution, see Prior
    distribution
Conditional probability, 57–63
Conservative play, 92, 251
Continuous random variable
    defined, 83
    examples, 157–184
    expected value, 161
    expected value of sums, 192
Convergence in distribution, 127, 210
Cooler, 36, 248–250
Correlation, 110, 194, 236
Covariance, 194, 236
Croak, Robert, 43, 118–119

Cumulative distribution function
  and convergence in distribution, 127
  defined, 83
  examples of, 84–85
  of Bernoulli random variable,
    138–139
  of geometric random variable, 143
  of Pareto random variable, 181

D

D'Angelo, Ryan, 107
Deal making, 89
Dedonno, Michael, 105, 269
De Melo, Moreira, 133
Demidov, Ivan, 78
Dependent events, 65–68
Dependent random variables, 191–195
De Silva, Upeshka, 55
Detterman, Douglas, 105, 269
DiCarlo, Salvatore, 80–81, 133–134
Discrete random variables,
    defined, 83
    examples, 84–85, 137–152
    expected value, 86
    expected value of sums, 192
    non-negative, 123
    variance of, 119
Disjoint events, *see* Mutually exclusive
      events
Distribution, 83
Distribution function, *see* Cumulative
      distribution function
Drift, in random walk, 228
Duhamel, Jonathan, 18
Durrett, Richard, 201, 221, 227, 270
Durrrr, *see* Dwan, Tom
Dwan, Tom, 217–220

E

Eastgate, Peter, 16, 153
Elezra, Eli, 52, 68, 120, 246–247
Elimination times, 124, 163
Elliott, Chris, 29
Equity
    and optimal strategies for simplified
        poker games, 167–170
    and skill in hold'em, 104–119
    and the fundamental theorem of
        poker, 204–207

and the indifference principle, 170–171
    defined, 94
    examples, 107–119
Esfandiari, Antonio, 43, 55, 131–132
Expected value
    and equity, 94–95
    and skill in hold'em, 106–107
    of a Bernoulli random variable, 139
    of a binomial random variable, 141
    of a continuous random variable, 161
    of a discrete random variable, 86
    of an exponential random variable, 174
    of a geometric random variable, 143–144
    of a negative binomial random
        variable, 146
    of a non-negative random variable,
        123, 143–144
    of a normal random variable, 179
    of a Poisson random variable, 150
    of a sum of random variables, 192
    of a uniform random variable, 164
    general usefulness in hold'em, 83, 85
Expectation, *see* Expected value
Exponential distribution, *see* Exponential
      random variable
Exponential random variable, 163–164,
      190–193
Express equity, *see* Equity
Express odds, 96

F

Farha, Sammy, 51, 68, 89–90
Featherstone, Paul, 16
Feller, William, 127, 221, 270
Ferguson, Chris, 168, 270
Ferguson, Tom, 168, 270
Fractal dimension, 183
Frequentist view of probability, 2, 13
Friedlander, Steven, 38
Fundamental theorem of poker, 204–207

G

Galfond, Phil, 52, 118–119, 217, 262
Gambler's ruin, *see* Risk of ruin
Game theory
    indifference principle, 170–171
    luck and skill, 105–108
    myopic rule, 85, 92, 101–102
    optimal strategy, 167–174

Gauss, Carl Friedrich, 177
Gaussian distribution, *see* Normal
    random variable
Geometric random variable, 143–146
Girshick, Meyer, 169
Gold, Jamie, 1, 8, 10, 12–13, 33, 35–36,
    51, 89–91, 98, 100–102, 262
Gordon, Phil, 74–77, 130, 132, 270
Gravity, 243–246, 249
Grey, David, 132
Greene, Dax, 11–12, 54
Greenstein, Barry, 43, 68, 89–90,
    131–132, 265

H

Hansen, Gus, 21, 36, 130–131
Harman, Jennifer, 68, 131–132, 194–195
Harrington, Dan, 17, 49, 74, 127, 147,
    199, 206–207, 214, 233, 270
Harrington's watch, 17, 147, 206,
Harris, Lance, 79
Hastings, Brian, 54
Heavy-tailed distribution, 181
Hellmuth, Phil, Jr., 16–17, 30–32, 50,
    63–64, 74–75, 103–104,
    108–110, 132, 176–177, 270
Help, in *R*, 245
High Stakes Poker
    Players' profits, 161, 219–220,
        234, 250
    Season 1, 89
    Season 2, 36, 96, 246
    Season 3, 68, 73, 90, 144
    Season 4, 51, 194, 211
    Season 5, 146, 219
    Season 6, 52
    Season 7, 40, 43, 118, 131, 213
Himmelbrand, Greg, 81, 236
Hinds, John Allen, 136
Holdem package for *R*, 243–248
Holz, Feder, 81
HSP, *see* High Stakes Poker

I

iid, *see* Independent and identically
    distributed
Implied equity, 106
Implied odds, 95–96
Inadmissible, 174

Independent and identically distributed
    (*iid*), 85, 141–146, 151, 160,
    174–181, 188–190, 200–221,
    229, 231, 240
Independent events, 57, 64–67, 77, 79,
    196, 236
Independent random variables,
    191–195, 201
Independent trials, 85, 87, 137, 140–143
Indifference equations, *see* Indifference
    principle
Indifference principle, 169–171, 189
Inter-elimination times, 124, 163
Inter-event times, 175, 181
Ivey, Phil, 73–74

J

Jackpot, 69, 151
Jacobson, Martin, 53, 56

K

Kalas, Kane, 55
Kassouf, Sheen, 229, 271
Kelly criterion, 228–229, 237
Kelly, John, Jr., 228, 270
Kim, Jae, 79, 103–104
Kim, Min, 92, 270
Klein, Bill, 43, 118–119

L

Lamphere, Adam, 81, 236
Lau, Ka Kwon, 133
Law of large numbers, 85, 118, 138,
    200–207, 218, 229, 234, 268
Lederer, Howard, 110–117
Library, 243
Likelihood function, 184
Lindgren, Erick, 145, 248
Lottery, 71
Luck in hold'em, 104–118, 133–136
Ly, Minh, 96, 128

M

Macdonald, Norm, 40
MAD, *see* Mean absolute deviation
Magriel, Paul, 233
Magriel's *M*, 233

Markov inequality, 123–124, 187
Martens, Jay, 21, 130–131
Matusow, Mike, 74–75, 90–91, 144
McBride, Kevin, 142
McEachern, Lon, 134
McKeehan, Joe, 13–14, 53–55, 82, 134
Mean absolute deviation, 119–120
Memorylessness property, 146, 176–177
Mercier, Jason, 40
Miller, Ed, 204, 271
Minieri, Dario, 110–117
Minkin, Kelly, 154
Mistake, 204–207
Moment generating function,
    definition and properties, 125–127
    examples, 126–127, 132
    of a binomial random variable, 142
    of a negative binomial random
        variable, 146
    of an exponential random variable, 123
    of a normal random variable, 189, 209
    of a Poisson random variable, 150, 154
    of a Uniform random variable, 165
    proof of the central limit
        theorem, 209
Montgomery, Scott, 16
Moreno, Andrew, 11–12, 54
Morgenstern, Oskar, 168–174, 189, 271
Moss, Johnny, 87
Movsesian, Julian, 40
Multiplication rule for independent
    events, 67
Multiplication rule of counting, 20
Multiplication rule, general, 66–67
Mutually exclusive events, 3–7
Myopic rule, 85, 92, 99–102

N

Neff, Peter, 29
Negative binomial random variable,
    146–147, 154
Negreanu, Daniel, 36, 63, 68–9, 80–1,
    96–7, 107–9, 120, 129, 135,
    194, 234, 246–50, 264
Neuville, Pierre, 52
Newhouse, Mark, 11, 42, 53, 79
Nguyen, Danny, 21–22
Nguyen, Scotty, 142
Normal distribution, *see* Normal random
    variable

Normal random variable,
    defined, 178
    properties, 178–180
    examples, 181, 187–188
    moment generating function, 189
    standard, 179, 181
    and central limit theorem, 208–209
    and confidence intervals, 215–220
    increments in a random walk, 222
    simulation and computation of
        probabilities, 240
Nuts, 52–53, 82, 107, 258

O

Odds against an event, 70–71, 77
Odds of an event, 70–71, 77
Odds ratio, *see* Odds against an event
Optimal play, *see* Optimal strategy
Optimal strategy, 167–174; *see also* Kelly
    criterion, Myopic rule
Or, meaning of, 3

P

Pareto random variable, 181–184, 188
Park, Robert, 81, 236
Payne, Antonio, 80
PDF, *see* Probability density function
Peat, David, 43, 132
Perez, Fernando, 107
Perkins, Bill, 40
Permutations, 22–25
Pescatori, Max, 131
Peters, David, 54
Phillips, Dennis, 78
Playing the board, 142–143, 258
PMF, *see* Probability mass function
Pnorm, 240, 250
Pocket pair,
    and Bernoulli random variables, 57, 84
    and binomial random variables, 93–94
    and expected value, 88, 140
    and four of a kind, 50
    and three of a kind or full house, 52
    and two pairs, 51
    probability of, 81, 84
    waiting time, 146, 187
Poisson process, 150, 175
Poisson random variable, 147–152
Poker After Dark, 16, 30, 110–111, 132

Posterior distribution, 184–190
Pot odds, 92–95, 132
Power, Chad, 81, 133–134
Prior distribution, 184–187
Probabilistic strategy, 17, 147, 206
Probability density function
  and expected value, 161
  and standard deviation, 161
  and variance, 161
  definition and examples, 157–160
  of exponential random variable, 174
  of normal random variable, 178
  of Pareto random variable, 181
  of tapered Pareto random variable, 183
  of the maximum of two independent
    uniforms, 166
  of the minimum of two random
    variables, 189
  of the product of two random
    variables, 188–189
  of uniform random variable, 164
Probability mass function
  and expected value, 86
  and standard deviation, 119
  and variance, 79
  definition and examples, 84–86
  of Bernoulli random variable, 138
  of binomial random variable, 140
  of geometric random variable, 143
  of negative binomial random
    variable, 146
  of Poisson random variable, 148
Proportional chip deal, 89
Pseudo-random, 241

R

R, 239–249
Rainbow flop, 51
Ramdin, Victor, 90–91, 95
Random variables, 83; *see also* Discrete
    random variables, Continuous
    random variables
Random walks, 221–227, 234, 236–237
Reflection principle, 222–225
Reiss, Ryan, 79
Relance, la, 168
Relative frequency histogram, 160–163,
    182, 211–213
Richey, Brett, 63
Risk of ruin, 222

Robertie, Bill, 17, 49, 74, 127, 147, 207,
    214, 233, 270
Roberts, Brian, 81, 236
Ross, Sheldon, 42, 231, 271
Rousso, Vanessa, 132
Ruffin, Phil, 43
Rule of Two, 10
Rule of Four, 10–14, 28–29, 54
Runif, 241–242
Running it multiple times, 194–195

S

Sample mean, 85, 200–220, 250
Sample space 3–4, 16–19
Sampling without replacement, 27, 65–66
Sampling with replacement, 65
Sarra, Tom, Jr., 50, 103
Schanbacher, Jack, 81, 236
Schwarz, Justin, 13–14, 53–54
SD, *see* Standard deviation
Selbst, Vanessa, 43
Self-similarity, 183
Shak, Beth, 63
SHAL, *see* Structured hand analysis
Sheikhan, Shaun, 68
Simple random walk, 221–227, 234, 236
Simplified versions of poker, 167–174
Simulation, 217, 222, 239–249
Sindelar, Dan, 56, 103
Skill in hold'em, 105–108
Sklansky, David, 204–207, 271
Smith, Dan, 55
Smith, Gavin, 16–17, 30
Standard deviation, 119–124, 139–142, 195,
    208–211, 218; *see also* Variance
Standard normal random variables, 179–181,
    209–210, 213–215, 240
Stern, Zvi, 55
Stirling's formula, 226
Strong law of large numbers, 201–202,
    218, 229
Structured hand analysis, 71–74
Sums of random variables, 191–196
Switch2, 247
Szentkuti, Shandor, 21–22

T

T distribution, 216
Tell, 78, 107

Test for rare condition, 72–73
Thorp, Edward, 229, 271
Timemachine, 245
Tonking, William, 56
Tourn1, 245
Trials, *see* Independent trials
Turyansky, Alex, 13–14, 53, 55, 79, 81–82

U

Uncorrelated random variables,
    194–197, 234
Uniform random variables, 164–168
Union of events, 6
Ury, Jack, 38

V

van Hoof, Jorryt, 53
van Opzeeland, Marcus, 80
van Patten, Vince, 131
Variance,
    of a Bernoulli random variable, 139
    of a binomial random variable, 141
    of a continuous random variable, 161
    of a discrete random variable, 119, 161
    of an exponential random variable, 175
    of a geometric random variable, 143
    of a negative binomial random
        variable, 154
    of a normal random variable, 179

of a Pareto random variable, 188
of a Poisson random variable, 150
of a sum of random variables, 193
of a uniform random variable, 165
Velador, Luis, 50
Venn diagram, 5–6, 61, 72
Voluntarily put chips in the pot,
    see VPIP
von Neumann, John, 168–169, 171, 174,
    189, 270–271
VPIP, 219–220

W

Wasicka, Paul, 1–5, 8–13, 28, 33–36,
    98–102, 131, 144
Waxman, Matt, 80
Weak law of large numbers, 200–201
Williams, David, 73
Wilson, Keith, 207
Winner-take-all tournaments, 85–89, 92,
    94, 99, 132, 204, 233
World Poker Tour, 21

X

Xena, 245–246

Z

Zeidman, Corey, 246–247